LORENZO RAUSA

CNC
BASIC PROGRAMMING COURSE

For lathes and milling machines

Copyright © cnc webschool ™ 2021
New York (U.S.A.)

info@cncwebschool.com
www.cncwebschool.com

All rights reserved under national law and international conventions.

First Edition: January 2021 (R1.14).

ISBN: 9798593504807

Table of contents

Foreword

1. **Course Introduction (1h)** 11
   1.1 Purpose 11
   1.2 Means 12
   1.3 Method 12
   1.4 Duration 13
   1.5 Lathe and mill 13
2. **Start-Up of the Training Software** 17
   2.1 Download of the SinuTrain Operate program 17
   2.2 Installation 18
   2.3 Creation of the lathe 19
   2.4 Download of the programs and import into SinuTrain 20
3. **From the Program to Graphic Simulation (2h)** 23
   3.1 Introduction 23
   3.2 Opening of the program 24
   3.3 Import of tool data 25
   3.4 Graphic definition of the blank part 27
   3.5 Start of simulation 27
   3.6 Program Selection for the Start of Production 30
4. **Name and Direction of the Axes (2h)** 31
   4.1 Layout of the axes according to ISO Standard 31
   4.2 X-axis and the Z-axis 32
   4.3 C-axis 33
   4.4 How to determine the positive motion of the rotating axes 35
   4.5 Y-axis 36
   4.6 B-axis 37
   4.7 A-axis 38
   4.8 Concept of interpolation 38

	4.9	Programming scheme 39
	4.10	Practical exercise 41
		4.10.1 Movement on the X- and Z-axes and angular orientation .. 41
		4.10.2 Calculation of the offset values 44
		4.10.3 Duplication, renaming and modification of a program 47

5. Programming Concepts (3h) 49
    5.1 Elements constituting a program 49
    5.2 Logical programming sequence 50
    5.3 Duration of the validity of an instruction 51
        5.3.1 Modal instructions and groups of origin 51
        5.3.2 Self-deleting instruction 53
    5.4 Instruction types 53
        5.4.1 Technological instructions 53
        5.4.2 Geometrical instructions 53
        5.4.3 Auxiliary instructions 'M' 54
    5.5 Complementary instructions 55
        5.5.1 Entering of comments 55
        5.5.2 Message display 55
    5.6 Automatic block numbering 56
    5.7 Practical exercise 57
        5.7.1 Program analysis 57
        5.7.2 Automatic block numbering 58
        5.7.3 Deletion of block numbers 60

6. Coordinate Systems (2h) 61
    6.1 Machine coordinate system (MCS) 61
        6.1.1 Machine zero point 62
        6.1.2 Characteristic point of the slide 63
    6.2 Workpiece coordinate system (WCS) 63
        6.2.1 G54 - G57: part zero point setting 64
        6.2.2 Tool offset 66
    6.3 Practical exercise 67
        6.3.1 Setting of the part zero point, use of MDA and JOG 67
        6.3.2 Display of the position in MCS and WCS 69
        6.3.3 Tool offset by touching the workpiece 72

7. Tool Call (2h) 73
    7.1 Introduction 73
    7.2 T: tool call and function M6 73
    7.3 D: tool offset values selection 74
    7.4 Correction of the tool wear 77

	7.5	Practical exercise... 78
		7.5.1 Creation of a tool 78
		7.5.2 Deletion of a tool 80
		7.5.3 Creation of a second tool corrector................................... 81
		7.5.4 Deletion of a second tool corrector................................... 82
		7.5.5 Mounting and removal of the tools in the turret 83
		7.5.6 Saving of tooling data (with license only) 83
8.	**Spindle Activation (2h)** 85
	8.1	Introduction.. 85
	8.2	SETMS: setting of the master spindle 87
	8.3	G97: Spindle rotation with constant number of revolutions 88
	8.4	G96: setting of constant cutting speed 89
	8.5	LIMS=: limitation of the maximum number of revolutions 90
	8.6	M3, M4, M5: setting of the rotation direction 91
	8.7	Instructions to a spindle which is not the master spindle.............. 91
	8.8	Choice of the functions G97, G96 and LIMS.............................. 92
	8.9	SPOS=: programming of the angular orientation 92
	8.10	Practical exercise... 93
		8.10.1 Calculation exercises 93
		8.10.2 Creation of a new main program...................................... 94
		8.10.3 Creation of a new subprogram 95
		8.10.4 Creation of a new folder 95
		8.10.5 Spindle angular orientation exercises review.................... 96
9.	**Setting of the Feed rate (1h)** 97
	9.1	Introduction.. 97
	9.2	G95: feed rate expressed in mm/rev... 97
	9.3	G94: speed in mm/min 97
	9.4	Calculation of the execution time for one pass 98
	9.5	Practical exercise... 99
		9.5.1 Calculation exercises 99
		9.5.2 Saving of folders and programs 100
10.	**Absolute and Incremental Coordinates (1h)** 101
	10.1	G90: absolute programming 101
	10.2	G91: incremental programming 103
	10.3	Mixed programming 104
	10.4	Diametrical or radial meaning of the values associated with X ... 104
	10.5	Practical exercise... 105
		10.5.1 Analysis of a program in absolute coordinates 105

    10.5.2 Analysis of a program in incremental coordinates 106
11. **Basic Functions to Define the Profile (3h)** 107
    11.1 G0: rapid movement 107
    11.2 G1: linear interpolation 108
    11.3 G33, G34, G35: threading in multiple passes 109
    11.4 G4: dwell function 110
    11.5 Practical exercise 111
        11.5.1 Example of the roughing of a profile 111
        11.5.2 Review of program comprehension 113
        11.5.3 Example of the programming of a threading 114
        11.5.4 Finishing of a profile 117
12. **Direct Programming of Rounds, Chamfers and Angles (2h)** 119
    12.1 Introduction 119
    12.2 RND= / RNDM=: execution of a round 119
    12.3 CHR= / CHF=: execution of a chamfer 121
    12.4 FRC= / FRCM: specific feed rate on chamfers and rounds 122
    12.5 ANG=: direction of a line defined by an angle 123
    12.6 Practical exercise 125
        12.6.1 Point to point and direct programming comparison 125
        12.6.2 Definition of the blank part data 127
        12.6.3 Programming of a workpiece 129
13. **Circular Interpolation (1h)** 131
    13.1 G2: circular interpolation in clockwise direction 131
    13.2 G3: circular interpolation in counterclockwise direction 132
    13.3 I, K, I=AC(…), K=AC(…): progr. of the radius center 133
    13.4 Definition of the working plane 135
    13.5 Practical exercise 136
        13.5.1 Programming of different radii 136
14. **First Test (2h)** 139
    14.1 Introduction to the test 139
    14.2 Tooling operations and cutting parameters 140
    14.3 Drawing of the part to create 141
    14.4 Copying & pasting of program parts 142
    14.5 Program correction 142
15. **Tool Radius Compensation (1h)** 143
    15.1 Introduction 143
    15.2 G42: Enabling with tool on right side of profile 148

15.3	G41: Enabling with tool on left side of profile	149
15.4	Enabling and disabling with G40	150
15.5	Practical exercise	151
	15.5.1 Program analysis	151
	15.5.2 Test of concept comprehension	154
15.6	Reloading of complete tool list	156

16. Three-Axis Mill: Programming (2h) 157

16.1	Introduction	157
16.2	Layout of the axes in a mill	157
16.3	The X-axis, the Y-axis and the Z-axis	159
16.4	C- and B-axis in machining centers	160
16.5	Five-axis interpolation	161
16.6	Programming scheme	162
16.7	Machine zero point and definition of the part zero point	162
16.8	TRANS/ATRANS: incremental shift of the part zero point	164
16.9	Position of the point controlled by the NC and tool geometry	165
16.10	Setting of tool rotation and feed rate	167

17. Practical Milling Exercise (3h) 169

17.1	Introduction	169
17.2	Creation of a three-axis mill (X, Y, Z)	169
17.3	Download of the programs and import into SinuTrain	170
17.4	Direct selection of the tools in the program	171
17.5	Graphic definition of the blank part	173
17.6	Drawing of the part to be created	177
17.7	Program, phase 1: execution of the external profile	178
17.8	Program, phase 2: roughing of the internal profile	179
17.9	Program, phase 3: finishing of the internal profile	180
17.10	Program, phase 4: execution of the holes	181
17.11	Program, phase 5: activation of graphic simulation	182

18. Climb Milling and Conventional Milling (2h) 183

18.1	Peripheral milling	183
	18.1.1 Introduction	183
	18.1.2 Chip section area	183
	18.1.3 Conventional milling: cutter and workpiece movement	185
	18.1.4 Conventional milling: cutting force distribution	185
	18.1.5 Climb milling: cutter and workpiece movement	187
	18.1.6 Climb milling: cutting force distribution	187

        18.1.7 Conclusions 188
  18.2 Face milling 189
        18.2.1 Introduction 189
        18.2.2 Chip section area 189
        18.2.3 Climb and conventional face milling 190
  18.3 Comparison of different types of milling 191
        18.3.1 Peripheral conventional milling 191
        18.3.2 Peripheral climb milling 191
        18.3.3 Face milling 191

19. Programming of Four Milled Parts (8h) 193
  19.1 Programming example with the use of TRANS 193
        19.1.1 Workpiece program 194
  19.2 Example for climb milling 196
        19.2.1 Workpiece program 197
  19.3 Programming example using polar coordinates 199
        19.3.1 Polar coordinates system 200
        19.3.2 Movement commands with polar coordinates 201
        19.3.3 Workpiece program 202
  19.4 Programming example with tapped holes 205
        19.4.1 Tapping functions 206
        19.4.2 Workpiece program 207

20. Second Test (2h) 211
  20.1 Introduction to the test 211
  20.2 Tooling operations and cutting parameters 212
  20.3 Drawing of the part to create 213
  20.4 Program correction 214

Foreword

This course is aimed at high school students and anyone who is approaching the world of machine tool programming for the first time. Teachers and professionals may explore more complex topics in the advanced course proposed in the book "CNC - 50 Hour Programming Course".
The text includes all the basic programming concepts and explains the "G-code" standard functions, i.e. the programming language at the basis of all numerical controls. The training and graphic simulation software offers free and unlimited access and faithfully reproduces a real numerical control on the computer.
The teaching method and the covered topics have been selected to spark the students' interest and curiosity in the study of the matter. The training course includes chapters and paragraphs both for theoretical and practical instruction. Paragraphs on theory contain drawings and diagrams that simplify the understanding of the text. The first practical experiences consist in the use of pre-drafted programs that give the students the opportunity to familiarize with the numeric control and its potential. Later you will learn how to write new programs with difficulty levels that are commensurate to the acquired experience.
The practical exercises are accompanied by the respective operating procedures that allow the students to learn on their own, reducing the need for the teacher's presence.

Periodical tests are offered in order to help the students and teachers assess progress achieved or to highlight the topics for review.

The total number of hours necessary for the understanding of the theoretical part and for carrying out the practical exercises will always be specified at the beginning of each chapter. The analyzed machines are a three-axis lathe (X, Z, C) with driven tools and a three-axis vertical mill (X, Y, Z).

All the programs used during the explanation and all the images contained in this book, which may be used at home or printed, viewed or projected in the classroom, may be downloaded from the website cncwebschool.com.

I would like to thank SIEMENS, DMG MORI and SANDVIK COROMANT for their constant support and dedication during the drafting of this book.

Lorenzo Rausa

1. Course Introduction (1h)
(Theory: 1h)

1.1 Purpose

The specific purpose of this course is to learn how to create the complete program of a part starting with its drawing. However, the course consists in the achievement of many small goals.

The correct programming of a numerically controlled lathe is the result from the combination of different types of knowledge; these are the small goals represented by the conclusion of every chapter.

Learning the meaning of codes and functions is not sufficient for the creation of a part, because it is also necessary to understand the types of operations that the machine is able to perform; it is furthermore important to set the tools and to calculate the cutting parameters, all useful steps in the creation of the sequence of the tooling operations to be programmed, i.e. the execution of the working cycle.

After completion of the part-program, it is necessary to know how to use the control panel of the machine, the operating sequences to enter the program, to edit it, to save it and to reload it, to offset the tools, to view the graphic simulation, test the program and start the machine in automatic cycle.

These goals can be achieved through commitment and allocation of study time, supported by the learning method provided in this book.

1.2 Means

This handbook, associated to the training and graphic simulation software SinuTrain Sinumerik Operate 4.8, provides the means to achieve the set-out goals.

Sinumerik Operate is the name of the latest operating system created by Siemens for the operation of its numeric controls.

The SinuTrain program faithfully replicates all the functions and screens of a numeric control on a computer screen, which, during this course, becomes an actual NC lathe.

Furthermore, a keyboard reproducing the control panel may be connected to the computer. Its use makes the simulated experience ever more similar to the programming of a real numerically controlled machine.

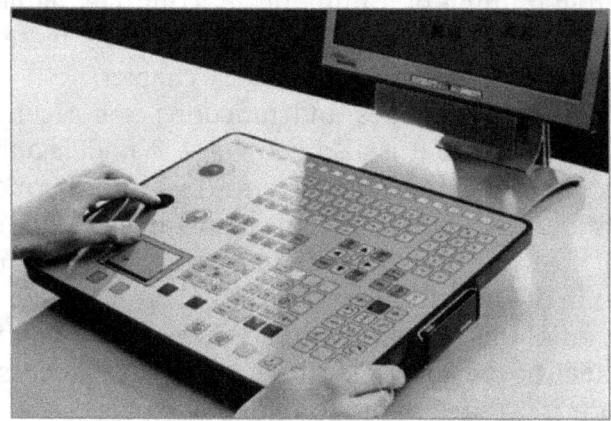

Fig. 1. Optional control panel for SinuTrain

1.3 Method

This course comprises chapters and paragraphs for both theoretical and practical learning. Paragraphs on theory contain drawings and diagrams that simplify the understanding of the text. The collection of all images and programs used during this course may be downloaded from the website cncwebschool.com in the TOOLS area. The first practical experiences consist in the utilization of pre-drafted programs, which are useful to the participants' initial understanding of the numeric control and its potential.

Later you will learn how to write new programs with difficulty levels that are commensurate to the acquired experience.

During the practical exercises, the reader is constantly guided by the respective operating procedures; in the learning phase on the use of a machine tool, the sequence of the buttons to push is always the most complex part to remember and the most boring part to take note of.

The learning method has been developed so that even beginners may complete the course and understand all the most complex functions and programming methods.

Periodical tests are offered in order to help the students and teachers assess progress achieved or to highlight the topics for review.

The course is based on the understanding of the 'ISO Standard' functions, i.e. the programming language at the basis of all numeric controls, whose commands have been encoded and, thereby, standardized by the International Organization for Standardization.

Knowledge of the ISO Code enables the programmer to operate on various numeric controls reducing the inconveniences caused by their inevitable differences.

1.4 Duration

The total number of hours necessary for the understanding of the theoretical part and for carrying out the practical exercises will always be specified at the beginning of each chapter.

The free license for the SinuTrain program has unlimited duration with regard to the examined machines (DEMO-Lathe and DEMO-Milling Machine). This will guarantee the completion of the course and the consolidation of the acquired knowledge.

1.5 Lathe and mill

The machines that are being examined are a three-axis lathe (X, Z, C) with driven tools and a three-axis vertical mill (X, Y, Z). They are single-channel machines and therefore only need one program to control the movements of all the axes.

This is the configuration of most machine tools in workshops all over the world and it is the ideal starting point to understand the operation and the programming of more complex machines equipped with more axes, more spindles or more channels.

The most important parts of the lathe are: the spindle, the turret and the numeric control.

The spindle holds the workpiece between three jaws that are normally powered by an oil hydraulic circuit. The spindle, also referred to as chuck, is able to place the center of the workpiece on its rotating axis.

The turret simulated in the training program offers twenty available positions which can all be equipped with fixed or driven tools.

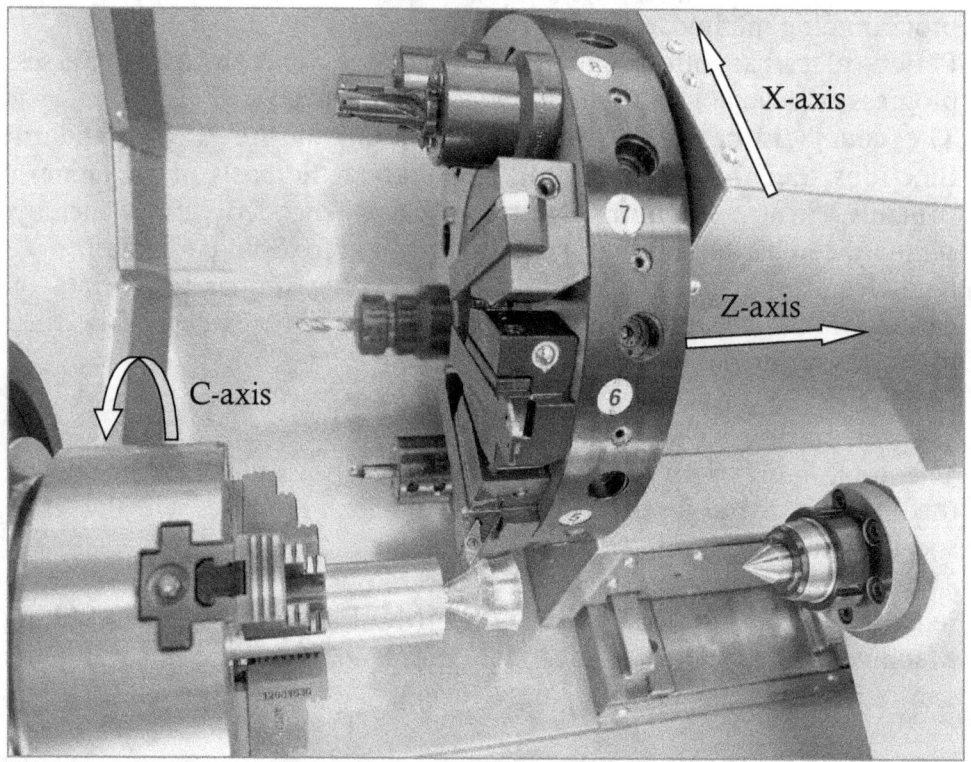

Fig. 2. Lathe with 3 axes and driven tools

The basic parts of the mill are the tool spindle, the workpiece table and the numeric control. The spindle rotates the tools normally mounted on tool holders with standard tapered attachments, the attachment and release of the tool holder are usually carried out by means of a robotized arm and a pneumatic suction and release system.
The manufacturer's choice of whether to move the workpiece table or the tool post does not affect the programming mode of the mill.
To simplify programming and program interchangeability, manufacturers are required to comply with ISO 841, which defines the nomenclature of axes and movements for CNC machines.

Fig. 2.1   Milling machine with 3 axes

The NC controlling both machines is produced by Siemens, Series 840D. This numeric control may be programmed in ISO standard language or via a conversational program called ShopTurn for lathes or ShopMill for mills.
We will not go into detail for this option as it is not in line with the goal to learn how to use the ISO functions.

2. Start-Up of the Training Software

2.1 Download of the SinuTrain Operate program

In order to proceed with the next step, the computer needs to be connected to the Internet.

Minimum PC requirements necessary for the installation and correct functioning of the SinuTrain program:

Hardware:	Processor 2 GHz, RAM 4 GB, Internet connection, USB data port
Disc Capacity:	Approx. 3.3 GB for full installation
Operating System:	Windows 7 SP1 (32 and 64 Bit) (no: Starter, Web Edition, Embedded) Windows 8.1 (32 and 64 Bit) (no: RT) Windows 10 (64 Bit) (no: Mobile, Mobile Enterpr.)
User Settings:	PC administrator rights required for installation and use
License:	The machines examined in the course (DEMO-Lathe and DEMO-Milling Machine) do not require any license

Fig. 3. Minimum PC requirements

Visit the website cncwebschool.com and access the TOOLS area to open the Siemens website. Register and take note of the access data you created to have them at your disposal for future access.

Username:	Password:
....	....

Fig. 4. Personal access data to Siemens website

After the 'login', activate the link for the download of the latest version of the simulation and training program named:

SinuTrain SINUMERIK Operate 4.8

A window for the *Download of the file* opens asking you if you want to save or open the compressed folder. Choose to save the folder and wait for the completion of the download.
Close the navigation window, select the downloaded folder with the pointer, push the right button of your mouse and select: *Extract all, Extract*. Wait for the completion of the operation.

2.2 Installation

Log in with a user with administrator rights. Open the folder and start the installation process with SETUP.EXE. You may be prompted to restart your computer; in this case restart your PC and resume the installation by clicking again on SETUP.EXE. Make sure your PC's firewall is deactivated. Make sure to select English when asked for the language settings of the numeric control.

On the next screen, three different programs are offered for installation.

SinuTrain Workbench is an application that allows you to create custom machines (optional choice).

SinuTrain is the name of the training program for the course (please select).

Automation License Manager is the program for the management of the licenses bought from Siemens. It is not necessary for the course but could be useful in the future (optional choice).

Wait for the completion of the installation of the selected programs.

2.3 Creation of the lathe

Start the program with the SinuTrain icon that has been created on the desktop of your PC, then push OK.

SinuTrain starts with an empty window which in the future will show the list of all machines created.

Now proceed with the creation of the lathe to be used during the course.

Click on "Use Template" to use one of the standard machines preconfigured within SinuTrain.

Now choose the machine type, change the name of the lathe, describe its basic features, set the size of the window which reproduces the machine video and the language you want to use. Enter the following information:

Template	DEMO-Lathe
Name:	LATHE: Programming Course
Description:	SP1-spindle (main spindle), X-axis (linear geometry axis), Z-axis (linear geometry axis), SP3-spindle (driven tool)
Resolution:	640x480 (or other resolution that best fits your screen)
Language:	English - English

Push CREATE.

The machine has been created and is now displayed on the starting page of the program.

Click on the newly created icon to start the lathe.

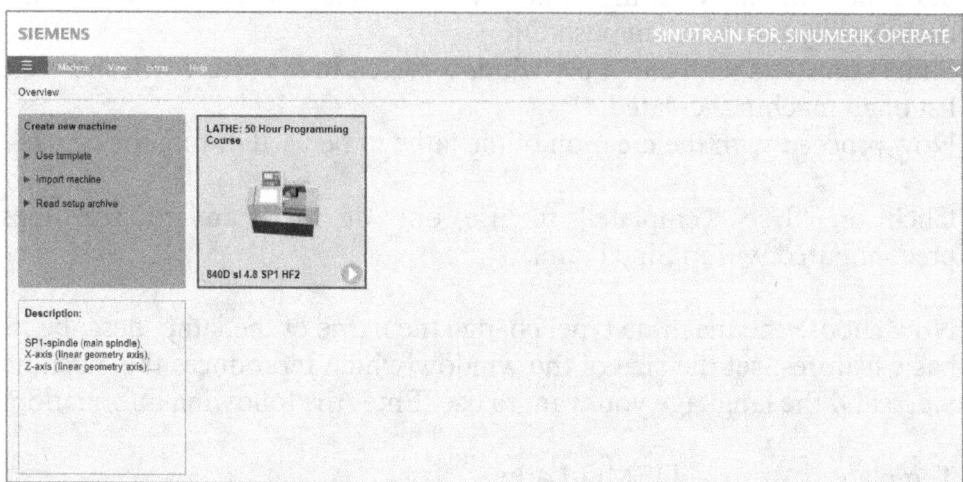

Fig. 5. Starting window of the simulation program

2.4 Download of the programs and import into SinuTrain

Open the website cncwebschool.com and access the TOOLS area to download the folder T3_PROG containing all the programs used during the course.

Select the compressed folder you just downloaded with the pointer, push the right button of your mouse and select: *Extract all*.

Now import the programs into the training software. Like a real machine tool, SinuTrain is able to upload and download data to and from an external memory connected via USB port.

Copy the folder T3_PROG onto an empty USB memory stick.

Select the newly created lathe and push START from the icons in the upper part of the screen.
On the control panel, click PROGRAM MANAGER.

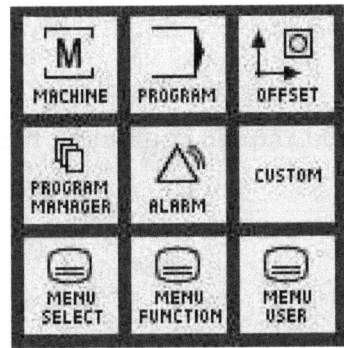

Fig. 6. Buttons for the selection of the operating environments

After selecting USB from the horizontal softkeys, the content of the USB memory is displayed on the screen.

Select the folder T3_PROG with the arrows and **push the yellow INPUT button to open it.**

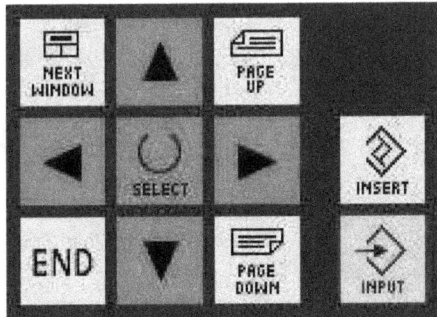

Fig. 7. Buttons for cursor movement and data entry

Now, place the orange selection bar on the first folder in T3_PROG named 01_EXERCISES.

Press MARK from the vertical softkeys and move down with the arrows until the whole content of the folder has been selected (or use the pointer).

Press COPY.

Push NC from the horizontal softkeys.

Select the folder WORKPIECES with the arrows and push PASTE from the vertical softkeys.

Now all the programs and the file containing the tooling data are ready for use.

3. From the Program to Graphic Simulation (2h)
(Practice: 2h)

3.1 Introduction

This exercise enables you to quickly understand all the contents of the course and consists in beginning from a program that has already been written in order to obtain a graphic simulation of the tooling operations and the finished solid of the part to be produced.

Furthermore, all the procedures for opening the program, defining the dimensions and the shape of the blank part, importing the tool data, starting the simulation and using the display options for the workpiece will be specified.

The part that you are about to create contains many of the tooling operations the machine is able to perform: roughing and finishing of the external profile, forming of the thread undercut, external threading with multiple passes, center drilling, axial drilling and tapping, radial drilling with angular orientation of the spindle, milling of the square with sides of 44 mm by interpolation of X-C and engraving of the inscription 'CNC' on the circumference.

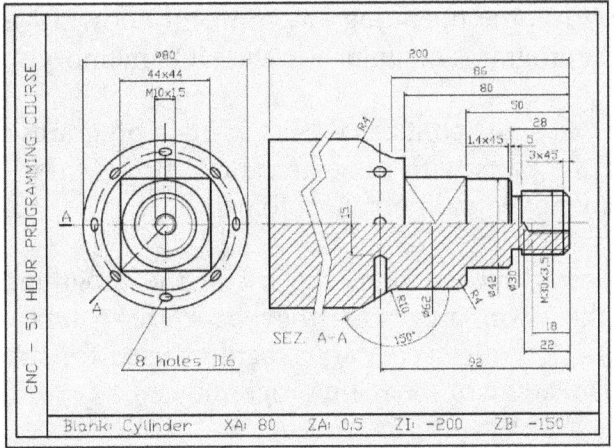

Fig. 8. Technical drawing of the introductory workpiece of the course

CNC – Basic Programming Course

3.2 Opening of the program

It should be remembered to follow this procedure every time when it is necessary to open a program in order to modify it or to execute the graphic simulation. Use the mouse to push the buttons displayed on the screen.
The control panel allows the operator to reach the program page either by pushing MENU SELECT and then PROGRAM MANAGER as shown in the following figure or by pushing PROGRAM MANAGER directly from the keypad section for the operating environments.

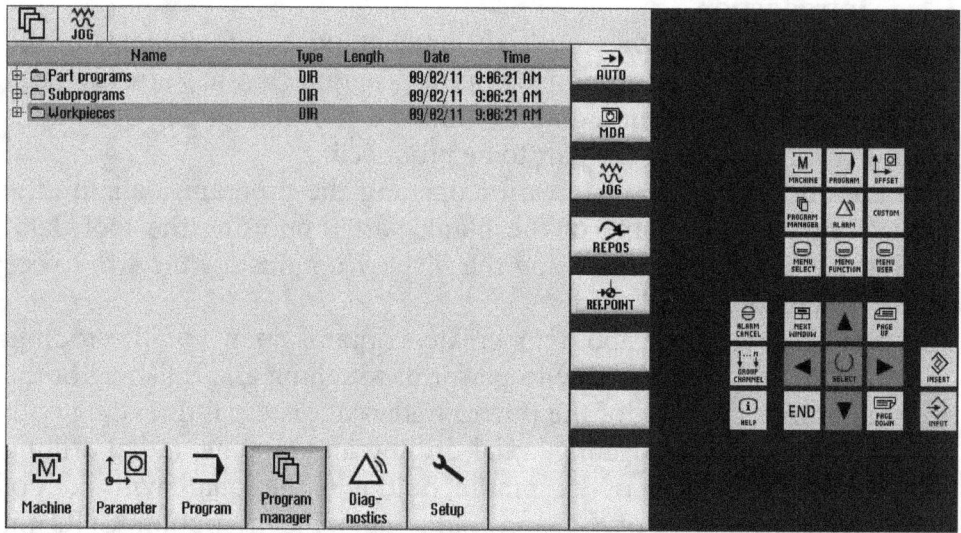

Fig. 9. Organization of the programs on the PROGRAM MANAGER page

Program Manager is the operating environment allowing you to manage and display the programs contained in the NC's memory.

The folder *Part programs* contains the list of main programs created for the production of the parts with the extension .MPF (Main Program File). This folder may not contain other subfolders.

The folder *Subprograms* contains the list of the programs that may be retrieved by the main programs; they have the extension .SPF (Sub Program File). They are secondary programs used for streamlining and simplifying the reading of the main programs.

The folder *Workpieces* may contain other subfolders and offers the possibility to organize the programs in different subgroups or to group all the programs for the production of a certain workpiece in one single folder, accordingly called 'workpiece folder'.
With the arrows, select the WORKPIECES folder, open it by pushing the yellow INPUT button, select the folder CHAP_03 and open it with INPUT.
In the folder you will find the program to be used in this chapter.
In order to open it select PRG_03_01 with the arrows and push INPUT.

3.3 Import of tool data

The data of the tools used by the program PRG_03_01 are gathered in a file that needs to be imported before starting the graphic simulation. Push PROGRAM MANAGER and then open the folder 01_EXERCISES, select the file TOOL_LIST with the arrows and push the yellow INPUT button. The NC recognizes that you want to load the tool data and proposes the following dialog box.

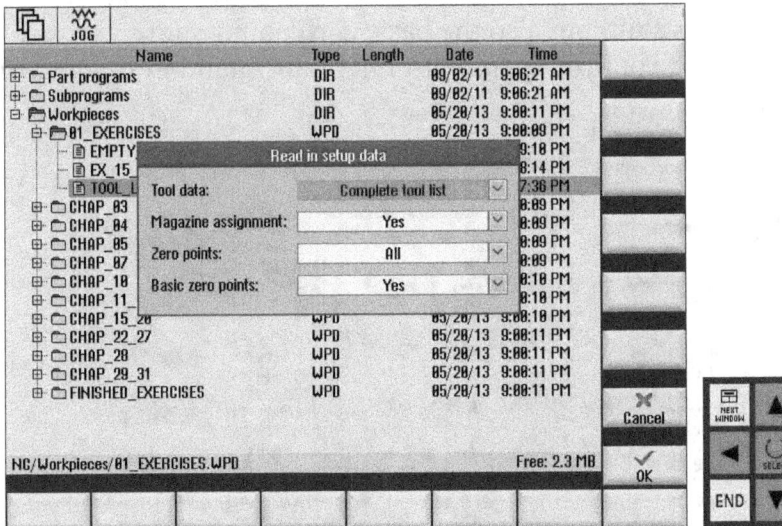
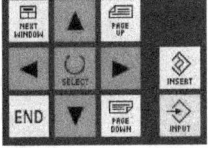

Fig. 10. Dialog box for the import of tooling data

Tool data: with the drop-down menu or the SELECT button placed at the center of the arrows, select: *Complete tool list*. This option overwrites the complete list of tools already defined in the machine.

If you select *No*, this means that you do not want to load the tool data but only the zero points and the basic zero points.
Move to the next items with the arrows.

Magazine assignment: with the drop-down menu or the SELECT button placed at the center of the arrows, select: *Yes*. This option loads all the tools into the same positions of the magazine where they were located when they were last saved. By selecting *No* the tools will be loaded into the positions following the 20 positions available in the magazine, allowing their future relocation on the turret using the LOAD and UNLOAD buttons.

Zero points: with the drop-down menu or the SELECT button placed at the center of the arrows, select: *All*. It is possible to upload the zero points of the axes. Select *No* if you do not want to load these data.

Basic zero points: with the drop-down menu or the SELECT button placed at the center of the arrows, select: *Yes*. This option allows to load not only the axes zero points shift, but also the basic zero points shift.
Then press OK and confirm again with OK your intention to overwrite the current data.

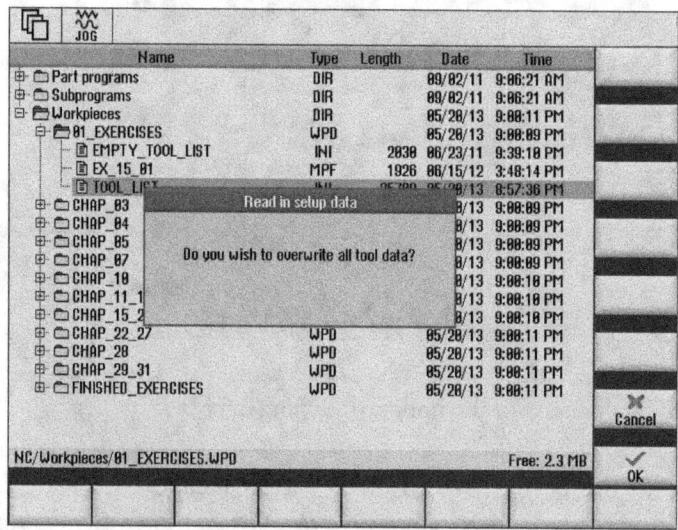

Fig. 11. Window for the confirmation and overwriting of the tool data

3.4 Graphic definition of the blank part

In order to obtain a graphic simulation faithfully replicating reality it is necessary to define the shape and the dimensions of the blank part to be worked on. In all the technical drawings analyzed during this course the data of the blank part are specified in the lower part of the drawing. They have the following meaning:

Blank part:	Shape of the blank part (e.g. cylinder)
XA:	External diameter of the blank part (e.g. 80 mm).
ZA:	Value of the machining allowance on the front face of the blank part (e.g. 0.5 mm).
ZI:	Length of the blank part. If by pushing SELECT you select ABSOLUTE (recommended), the length refers to the part zero point, if INCREMENTAL, the length refers to the front face of the part, machining allowance included.
ZB:	Extension of the face of the blank part from the jaws of the chuck. For the selection of absolute or incremental the same applies as for ZI.

Fig. 12. Description of the blank part dimensions

We will see later how to insert this information in the header of the program in the block: *WORKPIECE(,,, "CYLINDER", 0, 0.5, -200, -150, 80)*.

3.5 Start of simulation

Following the procedure described in paragraph 3.2, open the program PRG_03_01 with INPUT and press the horizontal icon in the lower right corner: SIMULATION (repeat this step two times at first program start).

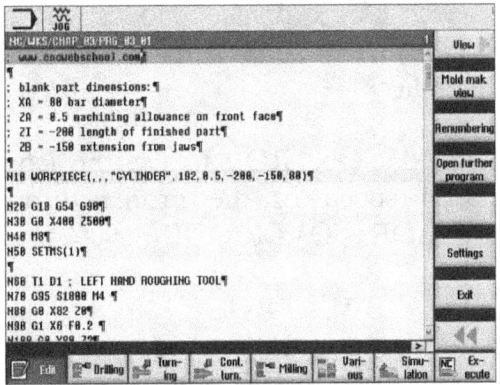

Fig. 13. Program opened and ready for simulation

Now use all the different display options:
SIDE VIEW, 3D VIEW, open FURTHER VIEWS to explore FACE VIEW and HALF CUT VIEW.

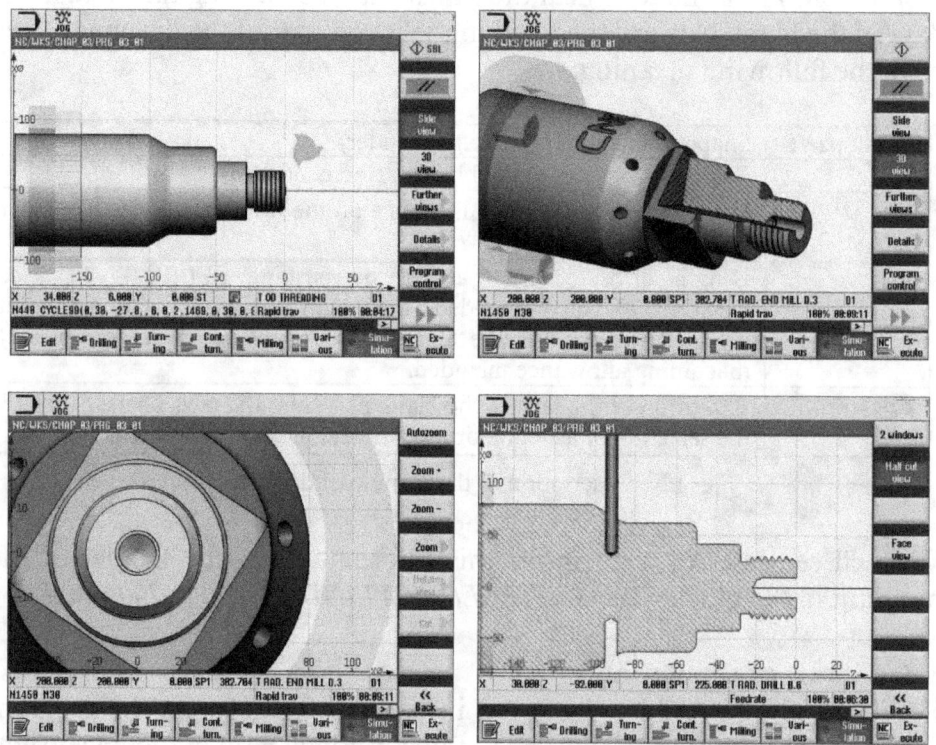

Fig. 14. Display options for the workpiece

Now discover, for each of the display modes, the additional features offered by the DETAILS icon in relation to ZOOM and CUTTING operations of the workpiece.

Then push PROGRAM CONTROL to use the icons which allow to change the execution speed of the profile through potentiometer management: OVERRIDE+, OVERRIDE-, 100% OVERRIDE.

After selecting 3D VIEW, PROGRAM CONTROL, push the OVERRIDE- icon and reduce the feed rate to 80%, then activate the SINGLE BLOCK icon to execute the program line by line.

This setting is very useful for the operator to associate the tool movements with the functions entered in the program, or to find the blocks causing an error in the execution of the profile.

Push BACK and start the execution of the program by pushing the green icon once and again, SBL stands for Single Block, which means that the single block mode is enabled.

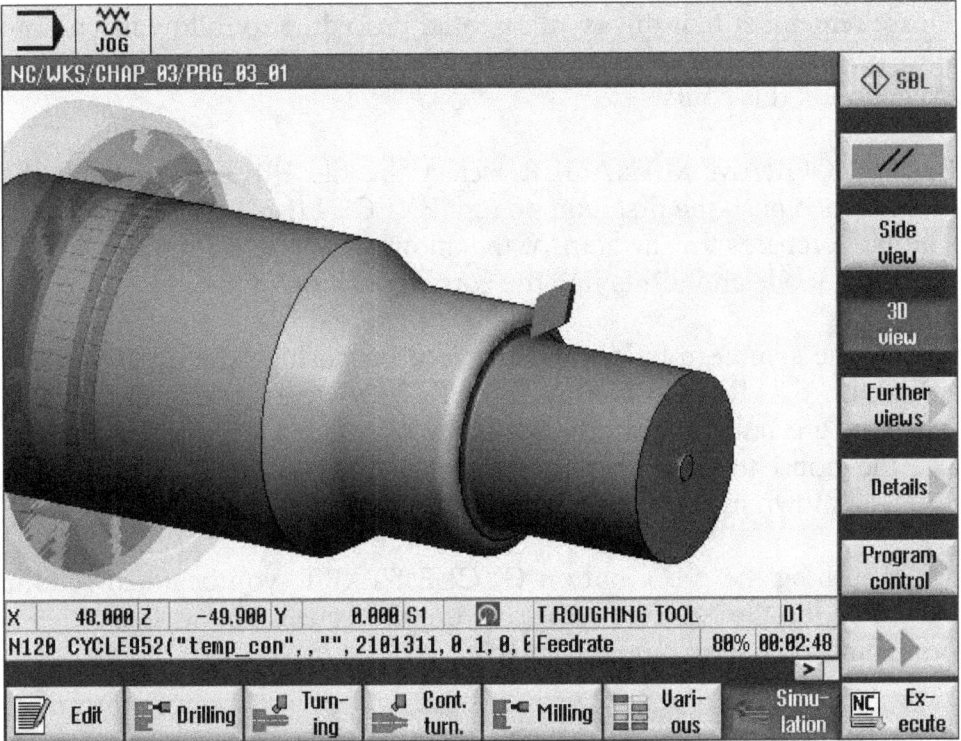

Fig. 15. Execution of the graphic simulation in single block mode

To exit the simulation and return to the program editor use the EDIT icon in the lower left corner of the NC screen.

Review the topics covered in this chapter and practice the use of the graphic simulation until you feel sufficiently confident.

3.6 Program Selection for the Start of Production

Before starting the production of a workpiece it is necessary to check the program by means of the graphic simulation, mount the tools on the turret, set the tool offset and then produce the first parts.

In order to determine which of the programs in the NC shall be carried out by pushing CYCLE START it is necessary to perform a "program selection".

Please remember that this is an essential procedure to follow in order to start the production of the machine, but it is not necessary during the execution of this course.

Push PROGRAM MANAGER, select the file PRG_03_01 with the arrows, then push the first icon on top: EXECUTE.

The NC prepares for the start of the automatic cycle setting the AUTO operating mode and displaying the current position of the axes.

Enable the spindle rotation and the feed by pushing the green buttons SPINDLE START and FEED START. They were successfully enabled when the green light above the buttons is on.

Use the mouse to rotate the potentiometer of the spindles and the feed to 100% as shown in the following figure.

After pushing the green button CYCLE START, you begin to see the execution of the selected program on the screen. Now the lathe is producing the part in automatic cycle.

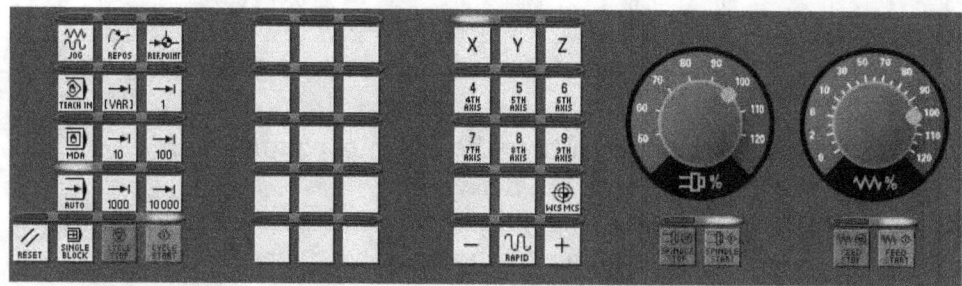

Fig. 16. Activation and setting of the potentiometers for the execution of the program in automatic cycle

4. Name and Direction of the Axes (2h)
(Theory: 1h, Practice: 1h)

4.1 Layout of the axes according to ISO Standard
Each axis is defined by the moving direction of the slide and is characterized by the capacity to interpolate with other axes in the machine. The presence of multiple axes in the machine means that the slide or slides move in various directions. ISO Standards have determined the name of every axis according to its direction and have defined their positive motion according to the sketch in the following figure:

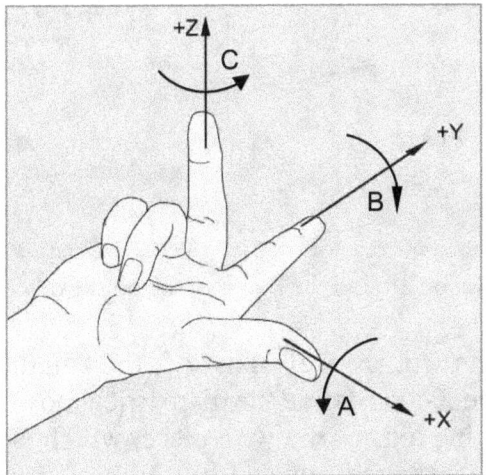

Fig. 17. Right-hand rule: definition of the axes and their positive motion according to ISO Standards. The positive motion is always determined in relation to the moving path of the tool on the workpiece.

This rule is called the "right-hand rule". The thumb represents the X-axis, the index finger the Y-axis and the middle finger the Z-axis. The same standard furthermore defines the names of the rotating axes. The axis rotating around X is called A, the axis rotating around Y is called B and the axis rotating around Z is called C.

4.2 X-axis and the Z-axis

In a lathe, the main axes are the X-axis and the Z-axis. They define the working plane X-Z. The movement of the tool in this plane together with the simultaneous rotation of the workpiece around the Z-axis leads to the creation of a solid by revolution. The Z-axis is therefore considered the axis which generates the part.

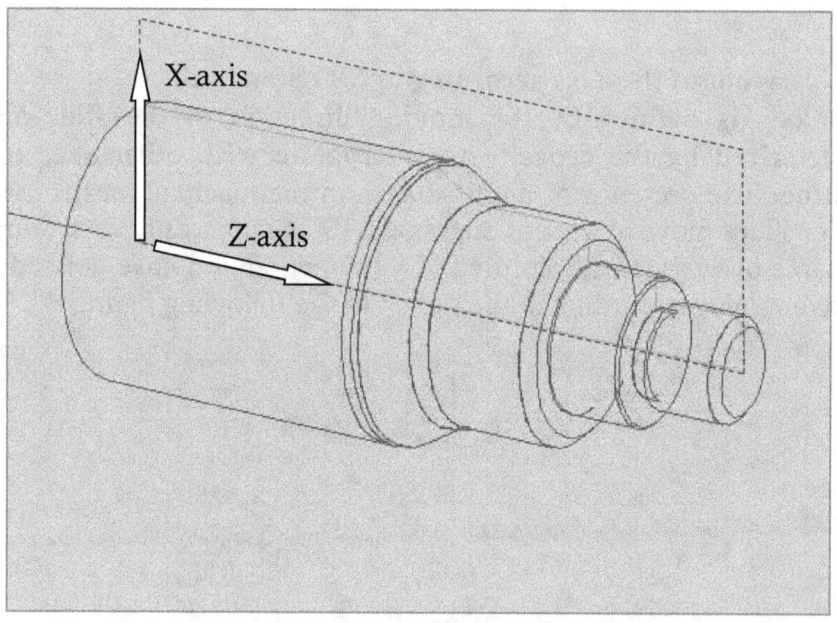

Fig. 18. Solid of revolution around the Z-axis of the profile described on the plane X-Z

The X-axis is the transversal axis of the diameters. The values programmed on the X-axis define the diametrical position of the tool with respect to the rotating axis of the workpiece, which is always considered equal to zero. The distance traveled by the tool, from one diametrical value to the other, corresponds to half of the difference between the two values; it is necessary to use this information during the programming of chamfers or grooves, which are often specified on the drawings with radial values.

The Z-axis is the longitudinal axis of the lengths. All values programmed on the Z-axis are real and refer to the part zero point which is normally located on the front face of the workpiece. The difference between two values in Z corresponds to the real distance traveled by the tool.

4.3 C-axis

The numeric control, thanks to its calculation functions, furthermore offers the opportunity to use the rotating axis of the spindle as an interpolating axis, i.e. it is capable of coordinating its movements on the basis of the movements of the other axes. The rotating axis of the spindle is always coaxial to the Z-axis and is therefore called the C-axis. With the C-axis, it is possible to perform milling and drilling operations, and its use is always associated with the presence of rotating tools in the machine which are called driven tools or live tools.

The C-axis can be used to orientate the spindle angularly in a way to perform radial holes along the direction of the X-axis or out-of-center holes along the direction of the Z-axis.

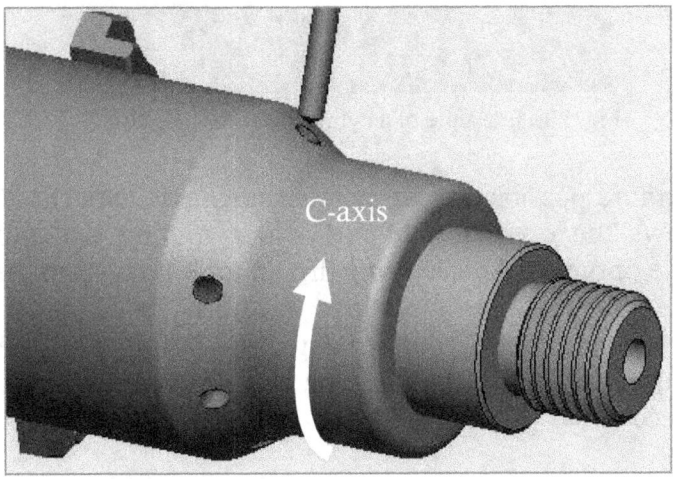

Fig. 19. Angular orientation of the spindle for the creation of radial holes

In lathes designed for processing very large workpieces, the stationing of the spindle in a certain angular position is guaranteed by the presence of a mechanical brake acting directly on a disc combined with the spindle itself.
In smaller machines the absence of the mechanical brake shows that the angular orientation and therefore the blocking of the spindle is obtained by keeping the motor of the spindle electrically active. The motor torque is the power used to contrast every movement caused by the tooling operations performed on the workpiece.

Another way of using the C-axis is to interpolate C with Z to perform milling operations on the workpiece surface. The programmed profile is described on a cylinder, and this is why this type of interpolation is called: cylindrical interpolation.

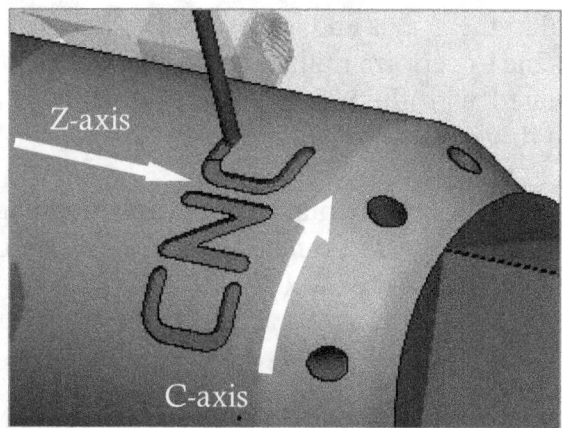

Fig. 20. Example of a cylindrical interpolation C-Z

It is furthermore possible to interpolate the C-axis with the X-axis. This option allows the execution of out-of-axis milling operations on the frontal plane of the workpiece using driven tools coaxial to the Z-axis. In this case the numeric control is able to transform the C-axis into a virtual Y-axis, i.e. an axis which is transversal to the workpiece and which allows for the creation of profiles of any type described on the frontal plane of the workpiece.

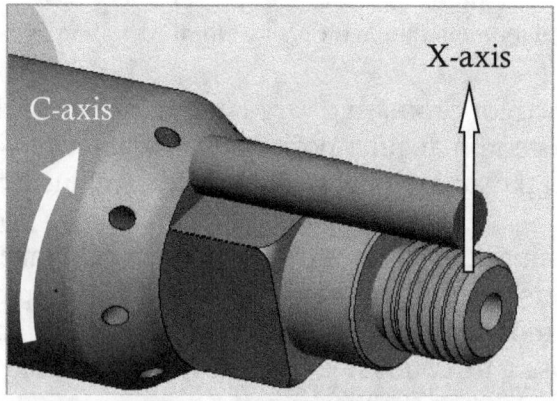

Fig. 21. Example of a frontal interpolation C-X

4.4 How to determine the positive motion of the rotating axes

The positive motion of the rotating axes is determined on the basis of the rule of the closing direction of the right hand.

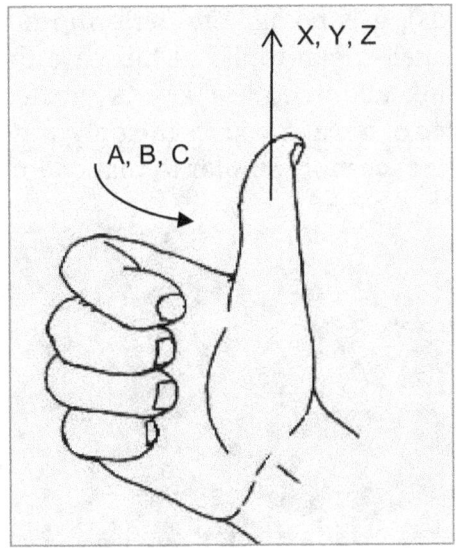

Fig. 22. Right-hand rule to determine the positive motion of the rotating axes

This rule determines the positive motion of the rotating axis; it is always referred to the movement of the tool.
If, on the contrary, the axis moves the workpiece, as in the case of the C-axis, the real movement of the spindle is in the opposite direction.

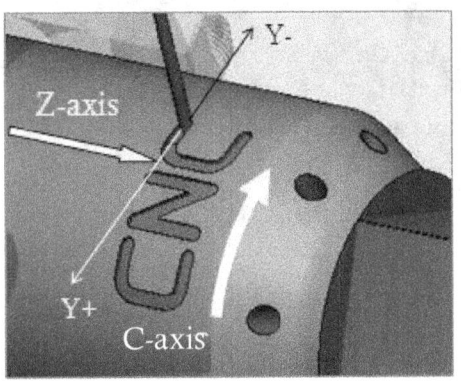 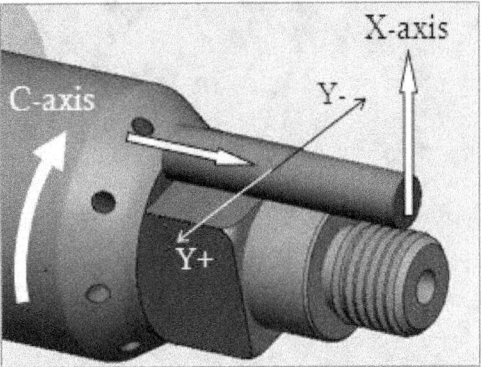

Fig. 23. C-axis positive programming direction and real movement of the workpiece

4.5 Y-axis

The Y-axis is present when the manufacturer of the lathe wants to offer a wider range of milling operations that the machine is able to perform. This axis is linked to the use of driven tools. Through the real Y-axis (and not virtual like the C-axis), it is possible to perform milling operations with flat bottom using radial driven tools. In this case the profile lies on the plane Y-Z. Just think about how a key is created and subsequently enlarged, the presence of a real Y-axis is the only way to create it, making the concept of a lathe ever more similar to that of a mill.

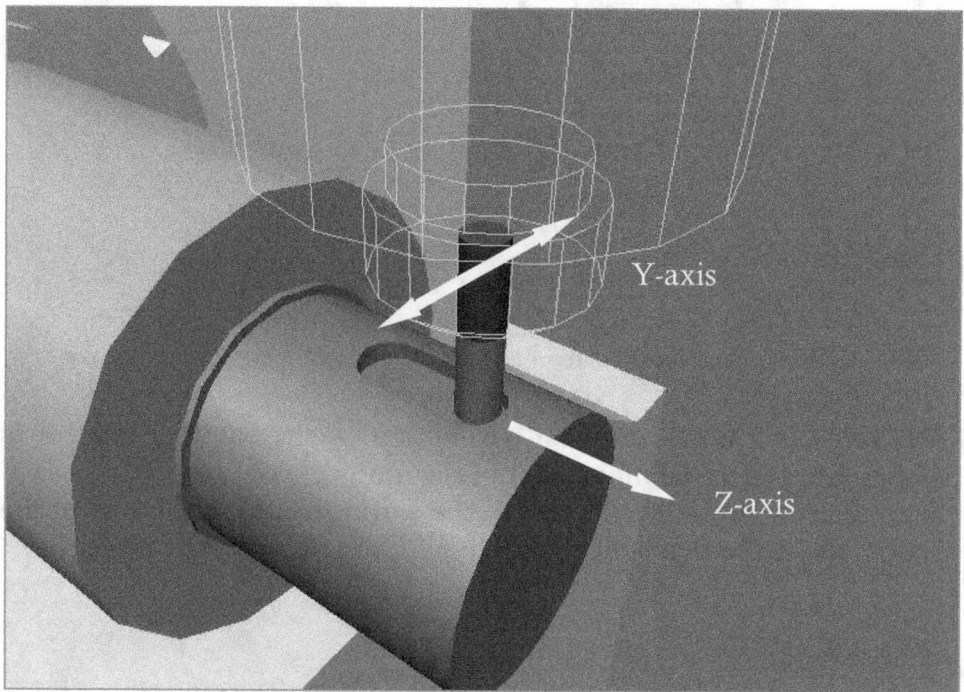

Fig. 24. Milling of a key using the real Y-axis

4.6 B-axis

The B-axis is the axis which rotates around the Y-axis; in lathes, it is used to perform inclined drilling and milling operations or gear hobbing with inclination with respect to the Z-axis. In this case, as for the Y-axis, the B-axis gives the lathe a higher flexibility of use for the milling operations.

The presence of the B-axis in lathes is restricted to a limited series of machines, usually equipped with an automatic tool magazine with tapered attachments instead of the more traditional rotating turrets.

The tool magazine, considered as a system for the deposit of tools outside of the working area, is usually a carousel or a chain type and allows for a much higher number of tools. It is therefore better suited to satisfy the complex tooling operations that this type of machine is able to perform.

The presence of a B-axis indicates that we are dealing with a machine which can not be simply defined as a lathe, but is in fact a universal machine able to perform turning operations as well as the most complex milling operations.

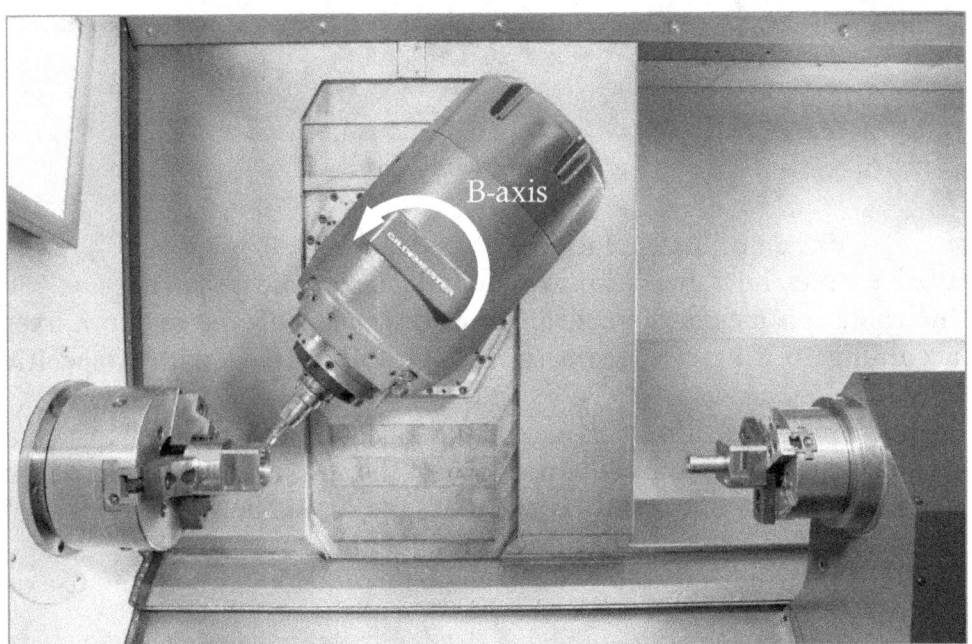

Fig. 25. Universal lathe with B-axis

4.7 A-axis

The A-axis is the axis rotating around X. The configuration of the majority of lathes does not include an A-axis as its functions are the same as the ones offered by a real Y-axis.

On the image below you see a machine with an A-axis which is able to perform rectilinear milling operations on the plane Y-Z thanks to the interpolation of the A- and Z-axes.

Fig. 26. Lathe with A-axis

4.8 Concept of interpolation

We have seen how the tool moves on independent Cartesian axes.

Interpolation means the coordinated movement of one or more axes according to a precise geometrical logic, performed with a specific speed.

Inclined lines, like those programmed to define a conical shape or a chamfer, are executed by interpolation of the X-axis with the Z-axis. In this case, the numeric control coordinates the simultaneous movement of the two axes so that the tool path is a straight line.

Milling on cylindrical surfaces is obtained by interpolation of the C-axis with the Z-axis.

Every time a working movement is programmed, there will be an interpolation; **the geometrical logic is defined by the current function enabled or set in the block, speed is defined by the programmed feed rate.**

4.9 Programming scheme

Most of the numerically controlled lathes use the programming scheme shown in the following figure.

The values associated with X are the diameter on which the tool or slide is located. Most of the time they are positive, X0 is on the rotating axis of the workpiece, negative values represent a position under the rotation center of the workpiece.

The values associated with Z express the longitudinal position of the tool or slide, Z0 is almost always defined on the front face of the workpiece, positive values mean that the position is outside of the workpiece, negative values mean that the tool is in operating phase or that it is in any case positioned beyond the front face of the workpiece.

The figure below shows the programming scheme which needs to be used for the assessment of the positive direction of the axes, in order to determine the clockwise and counterclockwise rotation of the circle arcs included in the profile of the workpiece, to define the right and left position of the tool with regard to the cutting direction and to assess the angle to be programmed in the event of inclined lines.

Standard ISO 841 defines a coordinate system of axes (fig. 17) where the positive direction is always referred to the movement of the tool.

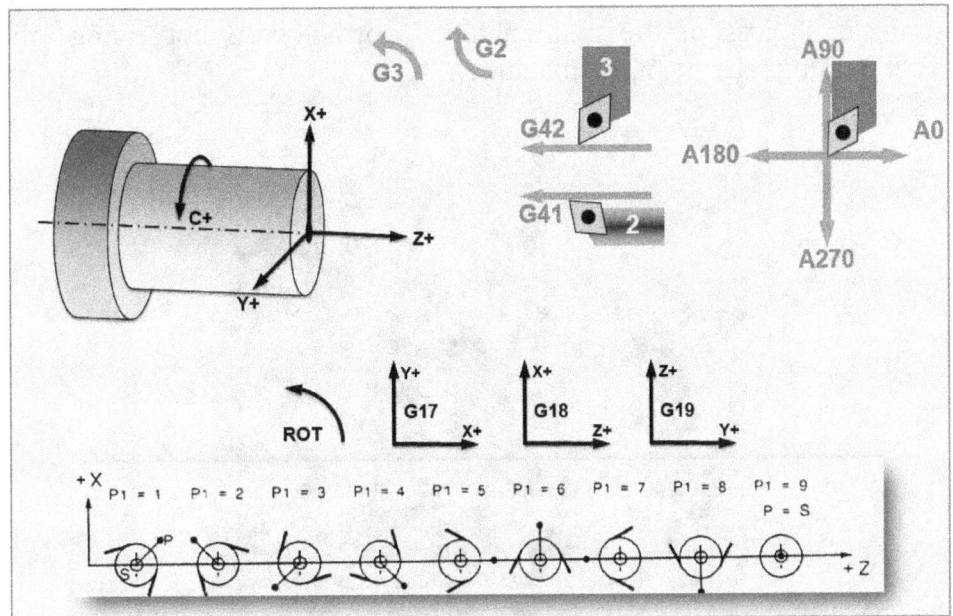

Fig. 27. Programming scheme

CNC – Basic Programming Course

The mechanical configuration of the analyzed lathe is the most widely used configuration and includes a turret moving on the workpiece which is fixed and held by the spindle.

Fig. 28. Traditional lathe where the tool moves on the workpiece

In some machines the real movement of the Z-axis or of the X-axis is made by the workpiece while the tool remains still. These features, due to technical choices of the manufacturer, normally do not change the programming scheme of the machine.

Fig. 29. Lathe with real movement of the Z-axis and of the X-axis on the workpiece

4.10 Practical exercise

4.10.1 Movement on the X- and Z-axes and angular orientation

This exercise aims at clarifying the meaning of the movement values to be programmed on the X- and Z-axis and of the angular values in relation to the orientation of the spindle.

A simple program containing movements on two axes is used to begin with. The executed operations are: turning along the Z-axis, two chamfers by interpolation of the Z-axis with the X-axis, four radial holes at 90 degrees obtained by angular orientation of the spindle.

The dimensions of the blank part are already in the program and the tools are those already used for the execution of the preceding workpiece.

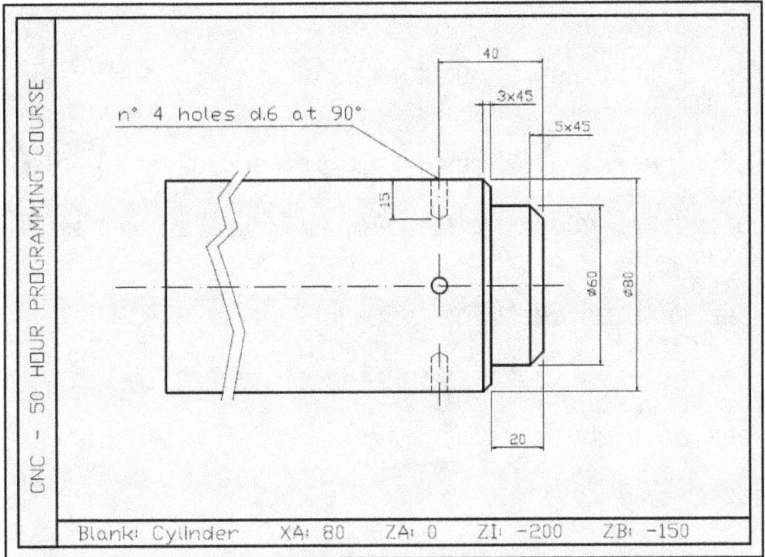

Fig. 30. Technical drawing created by the program PRG_04_01

Open the folder CHAP_04 and the program PRG_04_01 following the procedures specified in chapter 3.

Start the simulation by using the SIDE VIEW option and set the execution of the program to SINGLE BLOCK.

Below you will find the program for the creation of this part, try to proceed with the execution of the program while at the same time carefully reading the comments.

Check the correspondence between the drawing, the programmed value and the movement of the axis. It is not necessary to understand the whole program for this exercise.

```
N10 ; blank part dimensions:
N20 ; XA = 80 bar diameter
N30 ; ZA = 0 machining allowance on front face
N40 ; ZI = -200 length of finished part
N50 ; ZB = -150 extension from jaws
N60
N70 WORKPIECE(,,,"CYLINDER",0,0,-200,-150,80)

N80 G18 G54 G90
N90 G0 X400 Z500
N100 M8
N110 SETMS(1)

N120 T1 D1 ; LEFT HAND ROUGHING TOOL
N130 G95 S1800 M4
N140
N150 ; COMPREHENSION OF THE MOVEMENTS OF THE X AND Z AXES
N160
N170 ; DIAMETRAL MOVEMENT OF 20MM FROM X80 TO X60 MM
N180 G0 X80 Z0
N190 G0 X60
N200 ; THE REAL MOVEMENT OF THE TOOL IS 10 MM
N210
N220 ; LONGITUDINAL MOVEMENT OF 20MM FROM Z0 TO Z-20
N230 G1 Z-20 F0.2
N240 ; THE REAL MOVEMENT IS 20MM
N250
N260 G0 X62 Z0
N270
N280 ; FOR THE CREATION OF A CHAMFER 5x45
N290 ; THE MOVEMENT ON THE X-AXIS
N300 ; IS DOUBLE COMPARED TO THE MOVEMENT ON THE Z-AXIS
N310 G0 X50 ; INITIAL DIAMETER OF THE CHAMFER
N320 G1 X60 Z-5 ; FINAL VALUE IN X AND Z
N330
N340 G1 Z-20 ; LENGTH OF THE TURNING
N350 ; IF THE SECOND CHAMFER IS 3x45, THE INITIAL COORDINATE
N360 ; OF X IS 6MM BEFORE THE FINAL DIAMETER
N370 G1 X74 ; INITIAL DIAMETER
N380 G1 X80 Z-23 ; ARRIVAL POINT IN X AND Z OF THE CHAMFER
N390
N400 G0 X200
N410 G0 Z200
```

```
N420
N430 ; COMPREHENSION OF THE VALUES FOR THE ANGULAR ORIENTATION
OF THE SPINDLE
N440
N450 ; CREATION OF 4 HOLES, STAGGERED BY 90 DEGREES
N460 ; POSITION OF THE FIRST HOLE AT ZERO DEGREES
N470 ; ANGULAR ORIENTATION OF THE SPINDLE AT 0 DEGREES
N480 SPOS=0
N490 SETMS(3)
N500
N510 T8 D1; RIGHT HAND RADIAL DRILL D.6
N520 G95 S1200 M3
N530 G0 Z-40 ; LONGITUDINAL POSITION OF THE HOLES
N540 STR_HOLE1:
N550 G0 X84
N560 G1 X50 F0.1 ; FINAL DIAMETER OF THE DRILL
N570 G4 S2
N580 G0 X84
N590 END_HOLE1:
N600
N610 ; ANGULAR POSITION OF THE SECOND HOLE
N620 SPOS[1]=90
N630 REPEAT STR_HOLE1 END_HOLE1
N640
N650 ; ANGULAR POSITION OF THE THIRD HOLE
N660 SPOS[1]=180
N670 REPEAT STR_HOLE1 END_HOLE1
N680
N690 ; ANGULAR POSITION OF THE FOURTH HOLE
N700 SPOS[1]=270
N710 REPEAT STR_HOLE1 END_HOLE1
N720
N730 G0 X200
N740 G0 Z200
N750
N760 M30
```

4.10.2 Calculation of the offset values

This exercise assesses your understanding of the calculation method used to define the longitudinal, diametrical and angular positioning.

Now modify the positioning in X and Z and the values for the angular orientation of the spindle in order to create the following part. It is very similar to the previous part but with variations on the width of the chamfers, the length of the turning, the longitudinal position of the holes, the radial depth of the holes and the angular orientation of the holes.

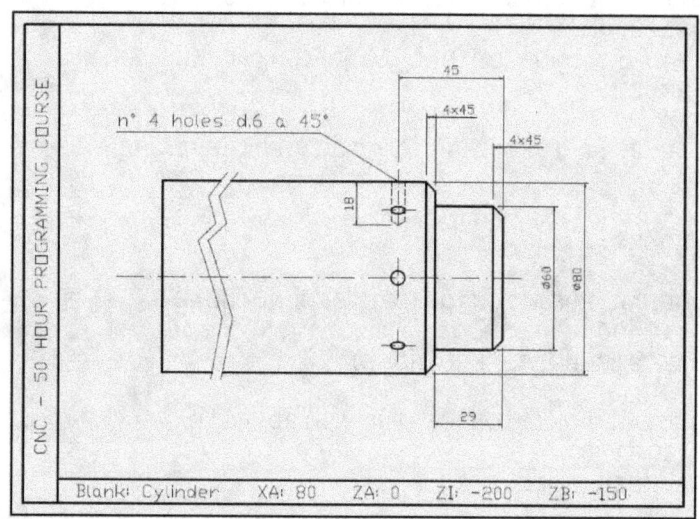

Fig. 31. Technical drawing of the part to create in program EX_04_01

The arrow (→) before the block number in the program below shows you where to enter the appropriate value. Then check with the graphic simulation.

```
N10  ; blank part dimensions:
N20  ; XA = 80 bar diameter
N30  ; ZA = 0 machining allowance on front face
N40  ; ZI = -200 length of finished part
N50  ; ZB = -150 extension from jaws
N60
N70  WORKPIECE(,,,"CYLINDER",0,0,-200,-150,80)

N80  G18 G54 G90
N90  G0 X400 Z500
N100 M8
N110 SETMS(1)
```

```
N120 T1 D1 ; LEFT HAND ROUGHING TOOL
N130 G95 S1800 M4
N140
N150 ; COMPREHENSION OF THE X AND Z AXES LINEAR MOVEMENTS
N160
N170 ; DIAMETRAL MOVEMENT OF 20MM FROM X80 TO X60 MM
N180 G0 X80 Z0
N190 G0 X60
N200 ; THE REAL MOVEMENT OF THE TOOL IS 10 MM
N210
→ N220 ; LONGITUDINAL MOVEMENT OF ……… MM FROM Z0 TO Z-………
→ N230 G1 Z-……… F0.2
→ N240 ; THE REAL MOVEMENT IS ……… MM
N250
N260 G0 X62 Z0
N270
→ N280 ; FOR THE CREATION OF A CHAMFER ……… x45
N290 ; THE MOVEMENT ON THE X-AXIS
N300 ; IS DOUBLE COMPARED TO THE MOVEMENT OF THE Z-AXIS
→ N310 G0 X……… ; INITIAL DIAMETER OF THE CHAMFER
→ N320 G1 X60 Z-……… ; FINAL VALUE IN Z
N330
→ N340 G1 Z-……… ; LENGTH OF THE TURNING
→ N350 ; IF THE SECOND CHAMFER IS ……… x45, THE INITIAL
→ N360 ; COORDINATE IN X IS……… MM BEFORE THE FINAL DIAMETER
→ N370 G1 X……… ; INITIAL DIAMETER
→ N380 G1 X80 Z-……… ; ARRIVAL POINT IN Z OF THE CHAMFER
N390
N400 G0 X200
N410 G0 Z200
N420
N430 ; COMPREHENSION OF THE VALUES FOR THE ANGULAR ORIENTATION OF THE SPINDLE
N440
→ N450 ; CREATION OF 4 HOLES, STAGGERED BY ……… DEGREES
N460 ; POSITION OF THE FIRST HOLE AT ZERO DEGREES
N470 ; ANGULAR ORIENTATION OF THE SPINDLE AT 0 DEGREES
N480 SPOS=0
N490 SETMS(3)
N500
N510 T8 D1; RIGHT HAND RADIAL DRILL D.6
N520 G95 S1200 M3
→ N530 G0 Z-……… ; LONGITUDINAL POSITION OF THE HOLES
N540 STR_HOLE1:
N550 G0 X84
→ N560 G1 X……… F0.1 ; FINAL DIAMETER OF THE DRILL
N570 G4 S2
```

```
N580 G0 X84
N590 END_HOLE1:
N600
N610 ; ANGULAR POSITION OF THE SECOND HOLE
→ N620 SPOS[1]=.........
N630 REPEAT STR_HOLE1 END_HOLE1
N640
N650 ; ANGULAR POSITION OF THE THIRD HOLE
→ N660 SPOS[1]=.........
N670 REPEAT STR_HOLE1 END_HOLE1
N680
N690 ; ANGULAR POSITION OF THE FOURTH HOLE
→ N700 SPOS[1]=.........
N710 REPEAT STR_HOLE1 END_HOLE1
N720
N730 G0 X200
N740 G0 Z200
N750
N760 M30
```

In order to correct the program wait until you have reached the end of this chapter.

4.10.3 Duplication, renaming and modification of a program

Normally, the programming of a new part does not start with an empty page. In order to reduce the time for starting the machine it is often decided to duplicate an existing program and modify it subsequently according to the new working cycle.

As the second program is very similar to an already existing program, it is duplicated and renamed. This procedure is very often used for numerically controlled machines.

Push PROGRAM MANAGER in order to access the program list, select the folder CHAP_04 with the arrows and open it with INPUT.

With the arrows, select the program to be duplicated: PRG_04_01. Press COPY. With the arrows, select the folder 01_EXERCISES where you want to insert the program and press PASTE.

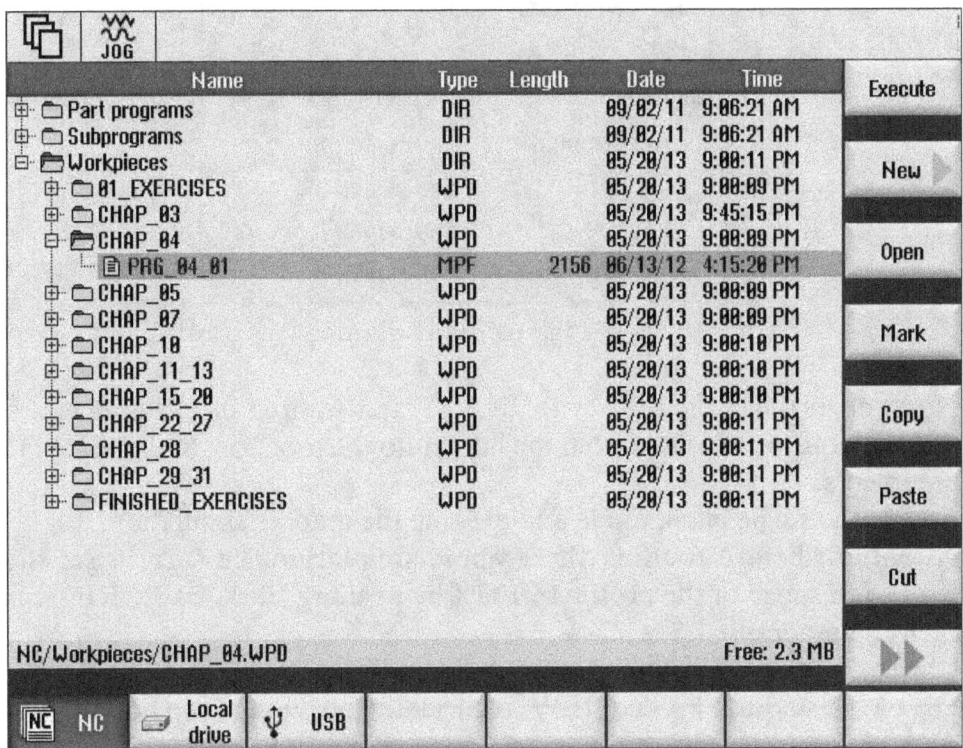

Fig. 32. Display of the PROGRAM MANAGER

The program is copied with the same name.
With the arrows, select the copied program and press NEXT to rename it.

Then press PROPERTIES. A window with a summary of the file properties opens. With your mouse, click on the name of the program and change it into EX_04_01, then press OK to confirm.

Fig. 33. File properties window

Now, open the copied file with INPUT and use the block numbers to transfer your conclusions from the book into the program, and update the comments.
Enable the single block mode and execute the graphic simulation.
Attention: before running the graphic simulation, be sure to set the execution speed of the profile to 100% by pushing 100% OVERRIDE as seen in paragraph 3.5.
Check the entered data and correct them if necessary. Use 3D VIEW, DETAILS and ROTATE VIEW to check the position of the holes.

Compare your program to the one in the folder FINISHED_EXERCISES named EX_04_01.

5. Programming Concepts (3h)
(Theory: 2h, Practice: 1h)

5.1 Elements constituting a program

The program consists of a sequence of instructions expressed by means of alphanumeric codes which give the machine all the necessary information in order to carry out a tooling operation.
The program is divided into a sequence of lines.
Every line is called "block" (e.g. G1 Z-20 G95 F0.1).
Every block contains one or more instructions defined as "words" (G1, Z-20, G95, F0.1).
Every word is made of a letter (G, Z, F) called "address", followed by a numeric value (1, -20, 95, 0.1).
The NC reads the program beginning with the first block, after the completion of the instructions therein it proceeds sequentially with the execution of the instructions entered in the following blocks until it arrives to the closing function of the program.

Block	Word	Word	Word	; Comment
Block	N10	G0	X20	; First block
Block	N20	G2	Z37	; Second block
Block	N30	G91	...	...
Block	N40	...	...	
Block	N50	M30	...	; End of program

Fig. 34. Name of the elements constituting the program

Every program has a name which is made of alphanumeric characters as well, there are no limits as far as its length is concerned, but only the first 24 characters can be displayed. The first two characters must be letters (or a letter with an underscore), followed by letters or numbers (e.g. _MPF100, SHAFT, SHAFT_2).

The sequence of the addresses programmed in a block does not impact the execution of the block itself. For a better understanding the following sequence is recommended:

N10 G... X... Y... Z... F... S... T... D... M...

The table below contains the first descriptions of some of the addresses which are most frequently used.

Address	Meaning
N	Address of block number
10	Block number
G	Preparatory function
X, Y, Z	Path information
F	Feed rate
S	Number of revolutions or cutting speed
T	Tool position
D	Number of tool corrector
M	Auxiliary function

Fig. 35. Meaning of some addresses

5.2 Logical programming sequence

When writing a program it is always recommendable to follow a precise logical sequence allowing not to forget any essential instruction.

The first element to be defined at the beginning of the program should be the **part zero point**, in other words the coordinate point X0 Z0 all the values saved in the program refer to; in a lathe, this point is often on the front face of the workpiece and on the rotating axis of the spindle.

Every single tooling operation is programmed as follows:
tool call, activation of the spindle, setting of the feed rate, rapid approach, execution of the tooling operation and disengagement from the workpiece with repositioning of the slide in the position where the tool is changed.

All the subsequent operations are programmed by repeating the same logical sequence.

5.3 Duration of the validity of an instruction

5.3.1 Modal instructions and groups of origin

Most instructions remain enabled in the blocks following the one where they were programmed; it is therefore not necessary to reprogram them until a function of the same type does not overwrite them. These functions are defined as modal functions.

The modal functions are deleted by functions belonging to the same group, i.e. functions which define similar instructions, but which contradict each other.

The function G1, which defines a working movement in linear interpolation, is deleted by the function G0, which belongs to the same group but defines a rapid movement.

The function G95, which sets the feed rate in millimeters per revolution, is deleted by the function G94, which belongs to the same group but sets the feed rate in millimeters per minute.

Below is the list of the most commonly used modal functions, subdivided by group of origin.

Name	Meaning
G0	Rapid traverse motion
G1	Linear interpolation
G2	Circular interpolation clockwise
G3	Circular interpolation counterclockwise
G33	Thread cutting with constant lead
G331	Rigid tapping
G332	Return (rigid tapping)
G34	Thread cutting with variable lead
G35	Thread with decreasing lead

Fig. 36. Group 1: Motion commands

Name	Meaning
G17	Plane selection 1st - 2nd geometry axis (X-Y)
G18	Plane selection 3rd - 1st geometry axis (Z-X)
G19	Plane selection 2nd - 3rd geometry axis (Y-Z)

Fig. 37. Group 6: Plane selection

Name	Meaning
G40	Deactivation of the tool radius compensation
G41	Activation of the tool radius compensation left of contour
G42	Activation of the tool radius compensation right of contour

Fig. 38. Group 7: Tool radius compensation

Name	Meaning
G500	Cancel all adjustable frames G54 - G57 if no value in G500
G54	Settable zero offset
G55	Settable zero offset
G56	Settable zero offset
G57	Settable zero offset

Fig. 39. Group 8: Settable zero offset (frame)

Name	Meaning
G60	Velocity reduction, precise stop
G64	Continuous path mode

Fig. 40. Group 10: Precise stop – continuous path mode

Name	Meaning
G70	Selects English units (inches and feet)
G71	Selects metric units (millimeter and meter)

Fig. 41. Group 13: Workpiece dimensioning inch/metric

Name	Meaning
G90	Absolute coordinate system
G91	Incremental coordinate system

Fig. 42. Group 14: Absolute/incremental coordinate system

Name	Meaning
G94	Linear feed mm/min or inch/min
G95	Rotational feed in mm/rev or inch/rev
G96	Constant cutting speed in m/min or feet/min
G97	Constant number of revolutions in rev./min

Fig. 43. Group 15: Feed rate and rotation type

5.3.2 Self-deleting instruction

Unlike the modal functions, **the self-deleting functions are valid only in the block where they are programmed**.

There are three self-deleting functions which are most widely used.

The first is G4, which sets the dwell time in seconds or revolutions; when the programmed dwell time ends, the function automatically disables and is not repeated in the next block.

The second is G9, which, entered in a certain block, sets an exact stop at the programmed arrival point.

The third is G53 and defines the machine coordinate system only in the block where it is programmed.

Name	Meaning
G4	Dwell time preset
G9	Exact stop only in the block where it is programmed
G53	Suppression of current frame

Fig. 44. Self-deleting instructions

5.4 Instruction types

The functions can also be grouped by type of command they set. Below are the most representative groups.

5.4.1 Technological instructions

Technological instructions are all those functions which define the cutting conditions.

Amongst these, we find the functions to select the position of the tool and define its corrector (s. fig. 35), those which set the cutting speed, the number of spindle revolutions and the feed rate of the tool (s. fig. 43).

5.4.2 Geometrical instructions

Geometrical instructions are linked to the definition of the reference system and to the tool path.

They define the type of tool path (fig. 36), the working planes (fig. 37), the activation of the automatic tool radius compensation (fig. 38), the reference system of the values programmed in the program (fig. 39), the exactness of the tool positioning (fig. 40 and 44), the type of measuring unit used (fig. 41) and the meaning of the numeric value (fig. 42).

5.4.3 Auxiliary instructions 'M'

Auxiliary instructions complete the information contained in a block.
With the 'M' functions it is for example possible to activate the cooling liquid, set the direction of the spindle rotation taking the back side of the spindle as reference, stop the program, set the end and other functions of the machine.

The meaning of most of the functions is decided by the manufacturer of the machine; it is therefore important to read the machine's manual in order to understand their meaning.

Below you will find a list of the instructions with fixed functionality used by most manufacturers.

Name	Meaning
M0	Programmed stop
M1	Optional stop activated by the control panel
M3	Spindle clockwise
M4	Spindle counterclockwise
M5	Spindle stop
M6	Tool change (if provided)
M8	Cooling liquid activation
M9	Cooling liquid stop
M30	End of program and return to beginning
M17	End of subroutine and return to main program
M40	Automatic gear change (when provided)
M41	Gear stage 1 (if provided)
M42	Gear stage 2 (if provided)
M43	Gear stage 3 (if provided)
M44	Gear stage 4 (if provided)
M45	Gear stage 5 (if provided)
M70	Spindle with transition to functioning as an axis

Fig. 45. Auxiliary or miscellaneous functions

5.5 Complementary instructions

In addition to the technological, geometrical and auxiliary functions there is a series of other commands which complete the programming. Below you will see how it is possible to enter comments, messages and the automatic numbering of the blocks in the program.

5.5.1 Entering of comments

In order to make the program clearer and more comprehensible, it is possible to enter comments.

Enter the comments at the end of the block and separate them from the block by means of a semicolon (;).

The comments are displayed together with the current block during the execution of the program.

```
N400 T1 D1 ;roughing tool
N410 X... Y...
N...
N500 T2 D1 ;finishing tool
N510 X... Y...
N...
```

5.5.2 Message display

The messages can be programmed to inform the operator of the tooling operation being currently executed.

The messages in the programs are created by writing the command MSG at the beginning of the block and then the message text between round brackets and quotation marks.

```
N400 MSG ("roughing") ;enabling of the message
N410 X... Y...
N...
N500 MSG () ;deletion of the message
```

The message text may have a maximum length of 124 characters, this will be displayed on two lines of 62 characters each.

5.6 Automatic block numbering

The block number is set by using the 'N' address, which identifies the position of the block in the program.

The number of the first block and the incrementation are defined by the programmer.

Before starting the graphic simulation it is always recommended to perform the automatic numbering of the blocks, which, in the event of a programming error, enables the NC to show the exact number of the line where the problem lies.

Normally, the number of the first block and the incremental value are both equal to ten; this allows to manually enter further blocks when modifying the program.

5.7 Practical exercise

5.7.1 Program analysis

Open the program 'PRG_05_01' in the folder 'CHAP_05' and look for all the elements of a program described in this chapter.
Start the graphic simulation in single block mode and analyze the movements of the tool according to the current program block with the help of the comments displayed.

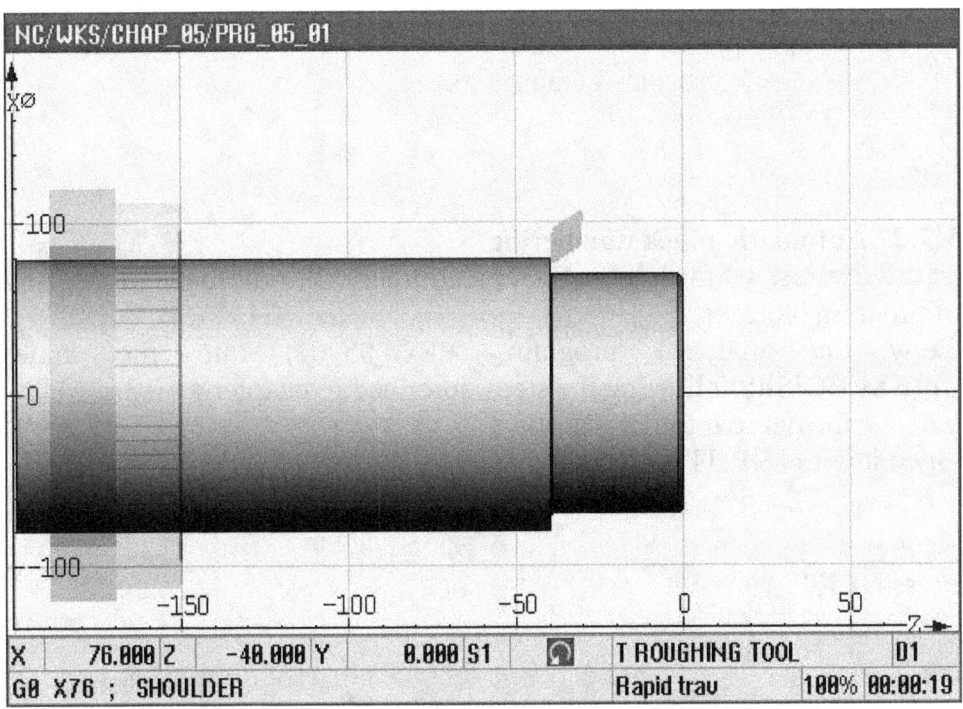

Fig. 46. Start of the simulation in order to analyze the program

```
; blank part dimensions:
; XA = 80 bar diameter
; ZA = 0 machining allowance on front face
; ZI = -200 length of finished part
; ZB = -150 extension from jaws

WORKPIECE(,,,"CYLINDER",0,0,-200,-150,80)

G18 G54 G90 ;G54 PART ZERO POINT SETTING
G0 X400 Z500
M8
```

```
SETMS(1)

T1 D1 ; LEFT HAND ROUGHING TOOL
G95 S1800 M4 ;SPINDLE ROTATION AND FEED RATE SETTING
F0.2 ; FEED RATE SETTING
G0 X82 Z0 ; APPROACH
G1 X-1; TOOLING OPERATION: FACING
G0 X66 Z2 ; DISENGAGEMENT
G0 Z0.5 ; APPROACH
G1 Z0 ; START OF TOOLING OPERATION
G1 X70 Z-2 ; EXTERNAL CHAMFER 2x45
G1 Z-40 ; TURNING
G0 X76 ; SHOULDER
G1 X82 Z-43 ; EXTERNAL CHAMFER 2x45
G0 X200 ; DISENGAGEMENT
G0 Z200
M30
```

5.7.2 Automatic block numbering

In this exercise you will learn the procedure for the automatic numbering of program blocks.

Copy the analyzed program (PRG_05_01) into the folder 01_EXERCISES following the steps described in chapter 4.10.3.

Change the name into EX_05_01.

Open it with INPUT.

Press NEXT:

Then RENUMBERING from the vertical softkeys:

The window for the block numbering appears:

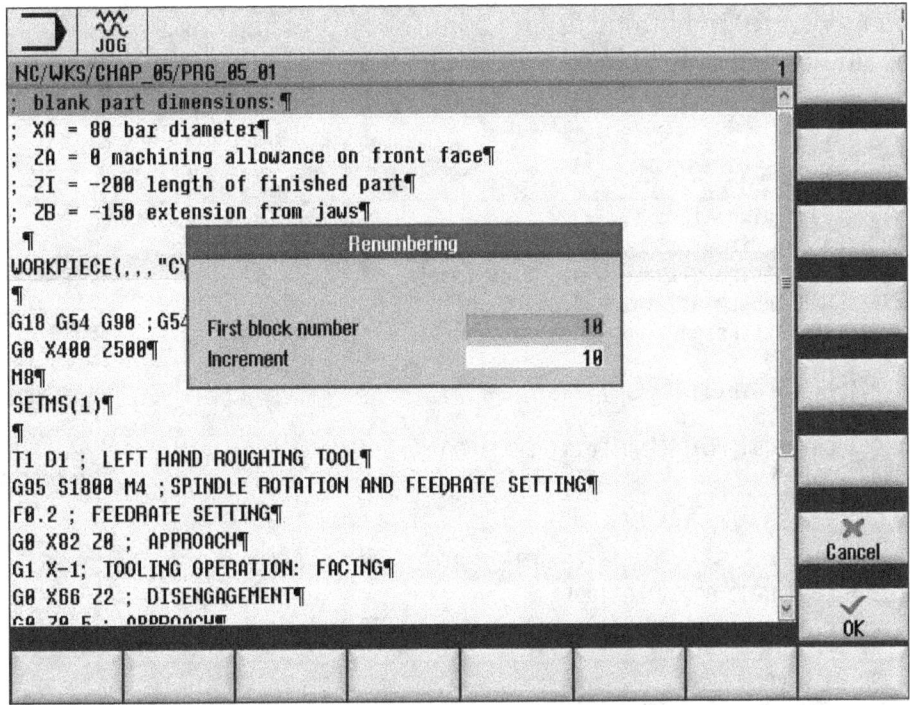

Fig. 47. Automatic block numbering screen

Enter the first number into the "First block number" field.
- Enter 10 and press INPUT.

In the "Increment" field, enter the incremental value for the numbering.
- Enter 10 and press INPUT.

- Press OK for the automatic numbering of the blocks.

Compare your program to the one in the folder FINISHED_EXERCISES named EX_05_01.

5.7.3 Deletion of block numbers

In order to delete block numbers go to the screen shown in fig. 47. Enter '0' (zero) in both fields and confirm by pressing OK. The numbering of the blocks will be deleted.

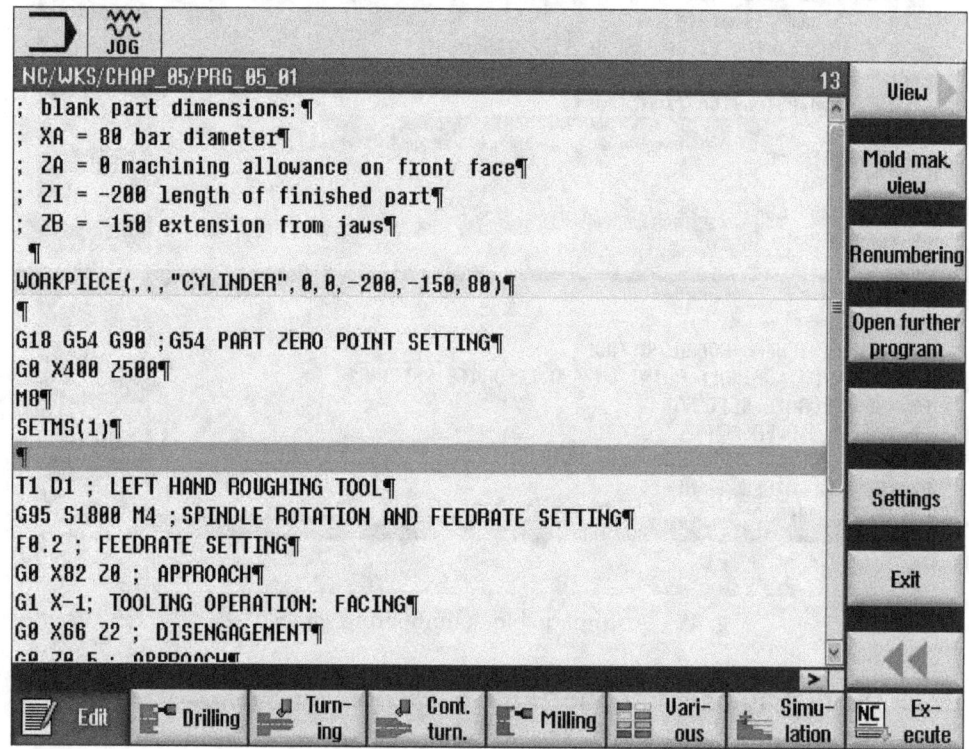

Fig. 48. Program without block numbers

6. Coordinate Systems (2h)
(Theory: 1h, Practice: 1h)

6.1 Machine coordinate system (MCS)

Every machine with a numeric control has a characteristic point to which the movements of the axes refer. This is called the **machine zero point**, i.e. the coordinate point X0, Z0, C0.

If there are no zero offsets enabled at the start of the machine, this is the only point the slides refer to.

All **slides have a characteristic point,** which is known to the NC; the coordinates displayed on the screen show the distance between the characteristic point of the slide and the machine zero point.

The reference system created is called machine coordinate system. This system is set by the manufacturer and can be modified by the operator through the program only.

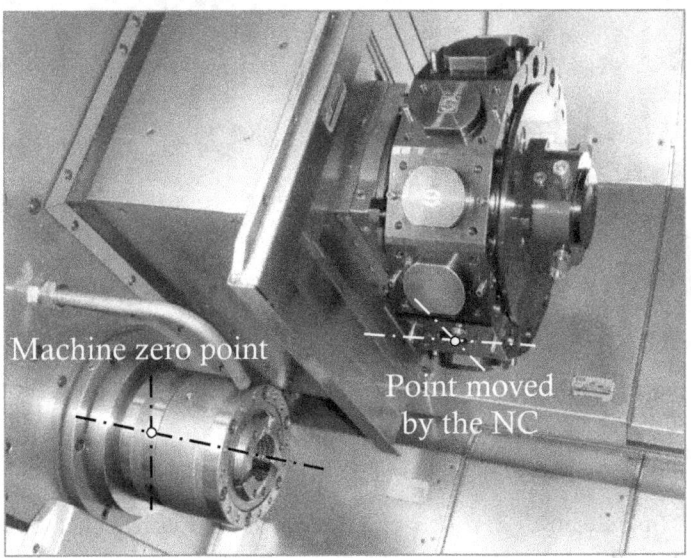

Fig. 49. Machine coordinate system: point moved by the NC referring to the machine zero point

The machine coordinate system is therefore independent of the length of the workpiece and the dimensions of the tool, it is not used during the processing of the workpiece but rather for the execution of safety positioning or general positioning in the working area.

6.1.1 Machine zero point

The machine zero point of a lathe is usually to be found on the face of the spindle nose and on the main rotating axis (fig. 50.1).

The spindle nose is the centering element upon which the different applications for holding the workpiece are mounted.

The position of the machine zero point is chosen by the manufacturer who prefers a point firmly connected to the base of the machine, independently of the dimensions of the devices used to hold the workpiece (fig. 50.2.3.4).

Fig. 50. 1: Spindle nose ; 2: chuck with three jaws ; 3: Elastic collet for external hold; 4: Elastic expansion collet for internal hold of the workpiece

6.1.2 Characteristic point of the slide

The NC actually does not move the whole slide but only a characteristic point thereof. Its position compared to the machine zero point is the one displayed in the machine coordinate system.

The point moved by the NC is placed by the manufacturer in a position of the slide which is logical and easy to identify.

Undoubtedly, one of the most logical and easiest to find turret positions is its center of rotation. Another point is the center of the attachment hole of the tool holder.

In the analyzed machine, as shown in fig. 49, the point moved by the NC lies at the center of the tool attachment hole (on the Z-axis) and on the external plane of the turret (on the X-axis).

The position of the point moved by the NC varies in different machines and it is very important for the operator to know where it is located; therefore always read the manufacturer's manual.

6.2 Workpiece coordinate system (WCS)

The machine coordinate system cannot be used for the definition of the tool path. **Every tooling operation is always programmed in the workpiece coordinate system, which you get by shifting the machine zero point onto the front face of the workpiece and the characteristic point of the slide onto the tool tip.**

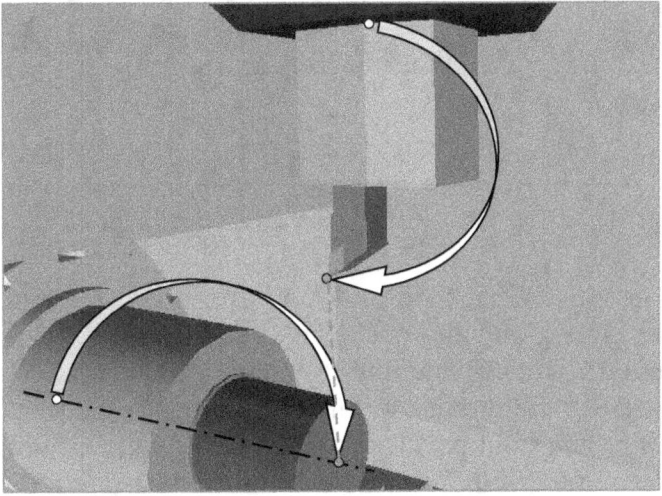

Fig. 51. Workpiece coordinate system: tool tip referring to the part zero point

6.2.1 G54 - G57: part zero point setting

The functions G54, G55, G56 and G57 offer the possibility to shift the machine zero point. The best position is on the front face of the workpiece. This new point is called the part zero point.

The part zero point is the coordinate point X0, Z0 all the values programmed after the enabling of one of the abovementioned functions refer to.

In a lathe, the position X0 normally is not changed, but left on the rotating axis of the workpiece.

Instead, the zero point along the Z-axis is shifted, **the offset value on the Z-axis corresponds to the distance between the face of the workpiece and the machine zero point.**

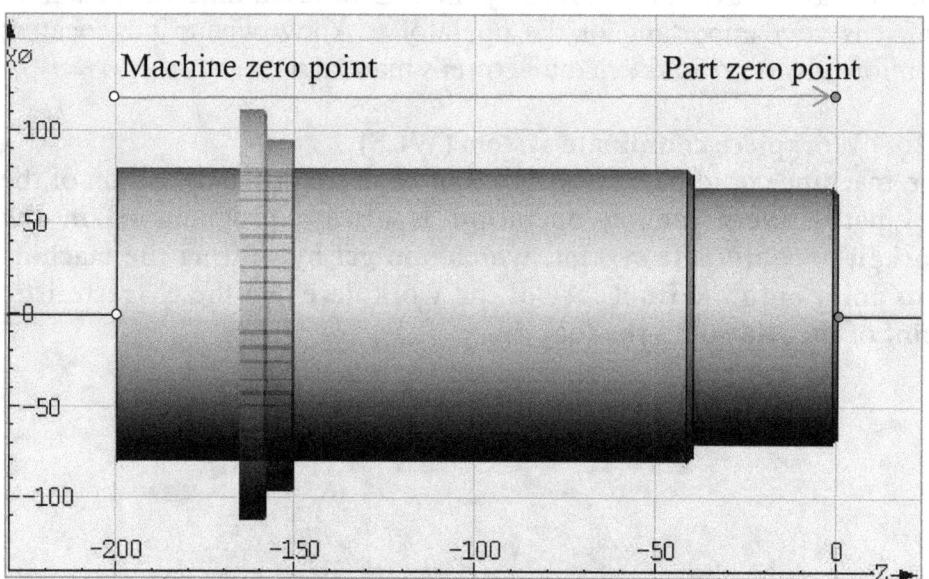

Fig. 52. Definition of the part zero point

The functions G54, G55, G56 and G57 are modal functions. They all belong to the same group and overwrite each other.

In every program, usually one (the most common being G54) function is used, or, if there are different clamping positions for the workpiece in the same cycle (first the front part is processed, the part is turned around and the back side is processed), two or more functions are programmed.

On the basis of the manufacturer's choices the function G54 may already be enabled at machine start and therefore it would not be necessary to enter it at the beginning of the program.

A table which is not part of the program contains the offset values which the operator will need to enter. **In order to get to this table press OFFSET on the control panel, then the horizontal softkey WORK OFFSET and then the vertical softkey G54... G57.**

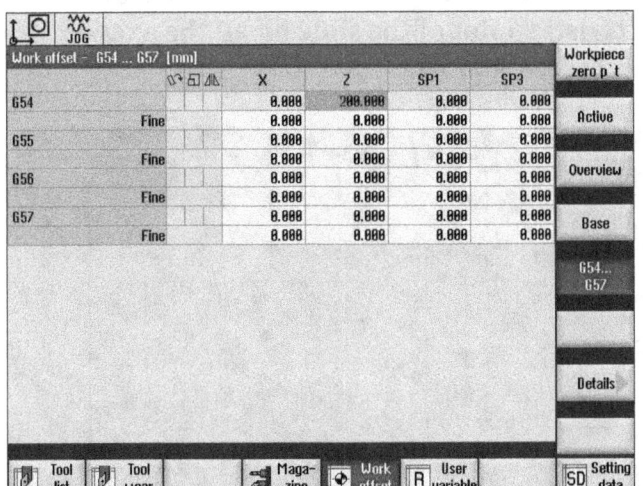

Fig. 53. Table of work offsets

In this table, next to the function, various columns can be found which offer the possibility to shift the machine zero point on all the axes set in the NC, including an angular value referring to the main spindle (SP1) and to the driven tools (SP3). In the second line, there is the possibility, for each function, to enter the correction to the main value.

It must be said that in most of the cases only the work offset on the Z-axis is used.

On the same page, by pressing the vertical softkey BASE, you will find the basic work offset, sometimes used by the operator to shift the machine zero point from the spindle nose to the front face of the jaws. The functions from G54 to G57 will later increment the 'basic' work offset up to the face of the workpiece.

The function G500, on the other hand, disables any work offset (when the manufacturer has not entered any value).

6.2.2 Tool offset

The tool path is described by programming the offsets of the tool tip in relation to the front face of the workpiece.

The tools used in the machine have different shapes and dimensions and this is something you have to tell your NC.

As we have already seen, the NC does not move a slide but rather a characteristic point thereof.

The position of the cutting edge is defined by the distance between the tool tip and the characteristic point of the slide on all the axes on which the slide moves (in this case X and Z).

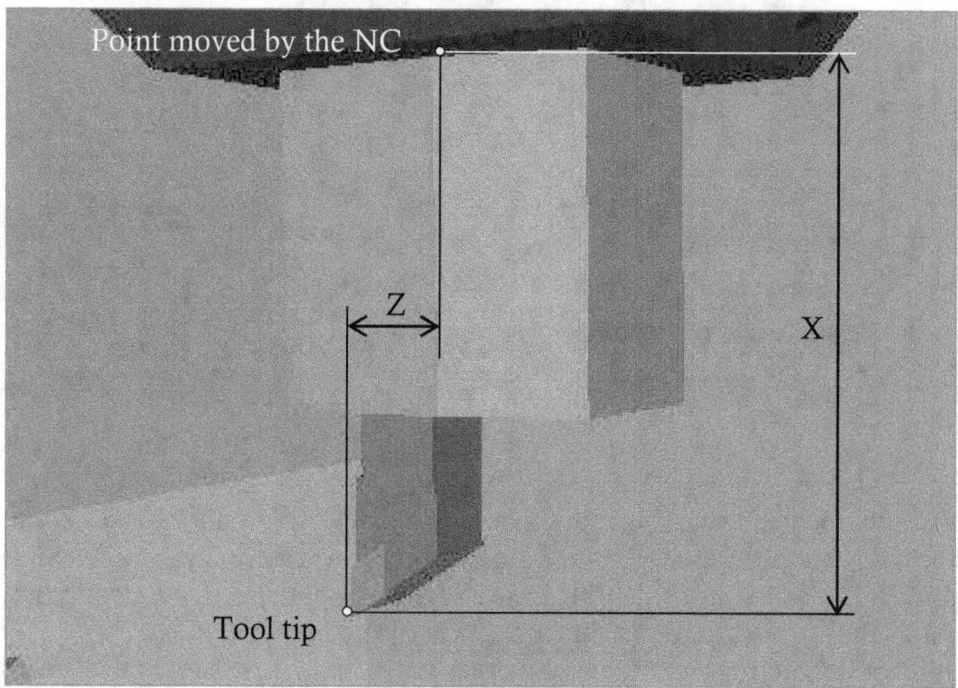

Fig. 54. Tool offset values

The offset values in X (for Siemens NCs) have radial and not diametrical value, i.e. they correspond to the real distance between the tool tip and the point moved by the NC.

These values are entered by the operator in the table for the tool offsets where furthermore the insert radius and other data in relation to the graphic description of the tool, necessary for correct simulation, are specified.

6.3 Practical exercise

6.3.1 Setting of the part zero point, use of MDA and JOG

Before setting the part zero point it is necessary to know how to use the operating mode MDA, i.e. the operating environment allowing you to enter data manually.

The MDA is often used to perform small programs, to call tools into position and to enable functions like work offsets.

In order to measure the distance between the face of the workpiece and the machine zero point, select a tool offset, touch the face of the workpiece and copy the current position of the tool tip into the table of the work offsets, next to the function G54, in the Z column.

On the control panel, press MDA under the NC screen: the page showing the position of the axes and the cursor blinking and ready for the manual data entry appear.

Now enter T1D1, which is the function to call the tool into position and to activate the corresponding offset values.

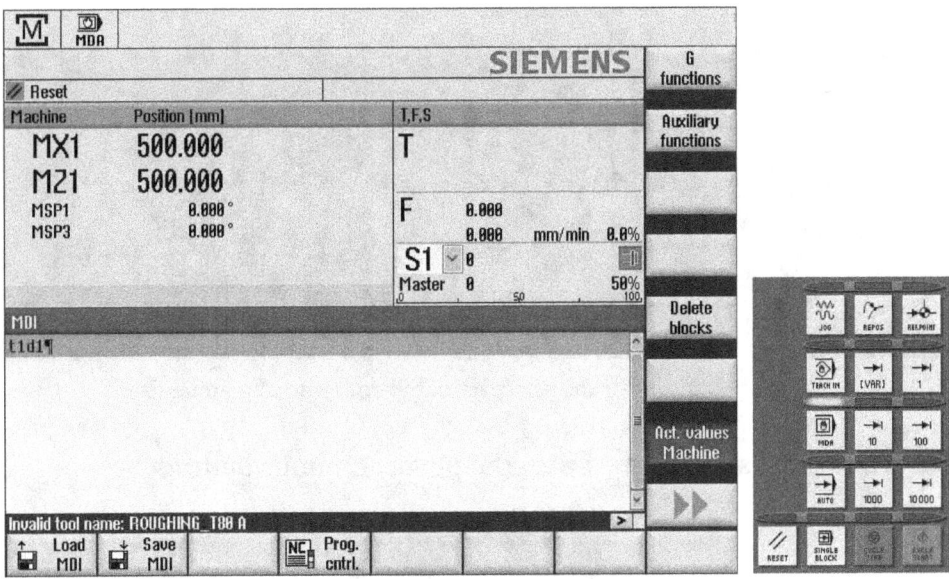

Fig. 55. Page for manual data entry

Press CYCLE START in order to execute the instructions. Now the turret has rotated to position the tool on the rotating axis of the spindle.

Press RESET to free the NC from the execution of the programmed block.

Now you are theoretically going to touch the face of the workpiece by moving the tool first in X and then in Z.

Push JOG.
Select the axis with which you want to move (X or Z).
Push the green button activating the spindle rotation.
Push the green button activating the feed.
With your mouse, bring the potentiometers (override) of the spindles and the feed to 100%.

Make sure that the LEDs show the activation of the buttons as shown in the following figure.

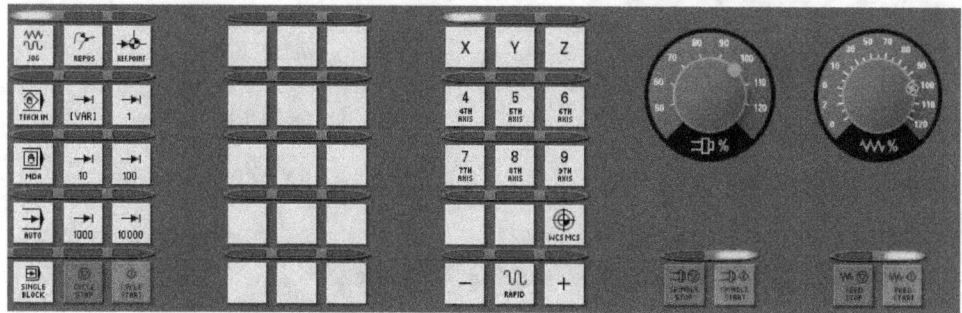

Fig. 56. Buttons for the selection of the continuous manual feed

Now move the selected axis with the plus and minus buttons.

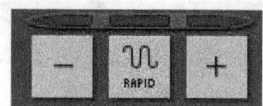

Suppose you are going to touch the face of the workpiece now. If the axes move too quickly, reduce the feed potentiometer exactly as you would do on a real machine.

Bring the values of the axes to X30 and Z200 and make sure to display the values in the WCS as described in the following paragraph.

6.3.2 Display of the position in MCS and WCS

The current position of the slide can be displayed in machine coordinates or workpiece coordinates.

By pushing **ACT. VALUES MACHINE** from the vertical softkeys the reference system changes.

Disable the button and make sure to be in the workpiece coordinate system. Now take the slide to a theoretical position to touch the face of the workpiece (e.g. Z200). **In order to arrive exactly at Z200 select one of the buttons which set the feed by incrementation, expressed on the keys in thousandths of a millimeter (10, for example, means one hundredth at a time).**

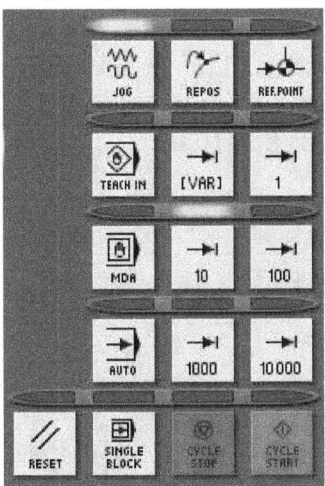

Fig. 57. Buttons for the selection of the manual feed by incrementation

Enter this value (200) into the table for the work offsets, next to the function used in the program (G54) and in the Z column (see fig. 53).

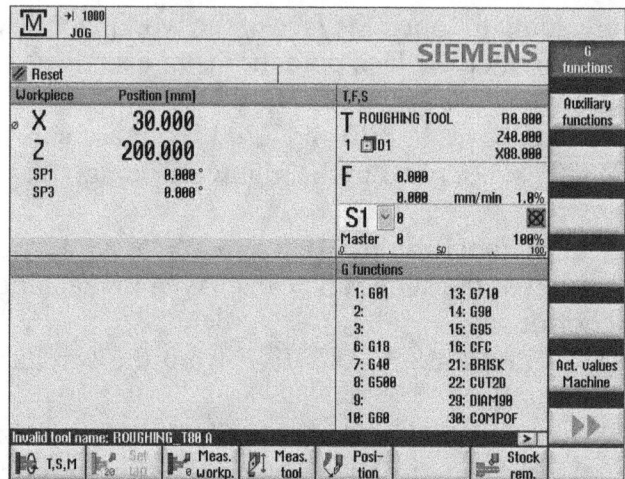

Fig. 58. Touch-off position of the part face in the workpiece coordinate system

Press MACHINE on the control panel to go back to the position of the axes. Press again ACT. VALUES MACHINE and select the machine coordinate system. The first difference you detect is that the position of X is expressed in a radial manner and not in a diametrical manner as in the workpiece coordinate system. The second difference is that the position in X and Z varies exactly by the tool offset value contained in the geometry table (X=(30/2)+88=103).

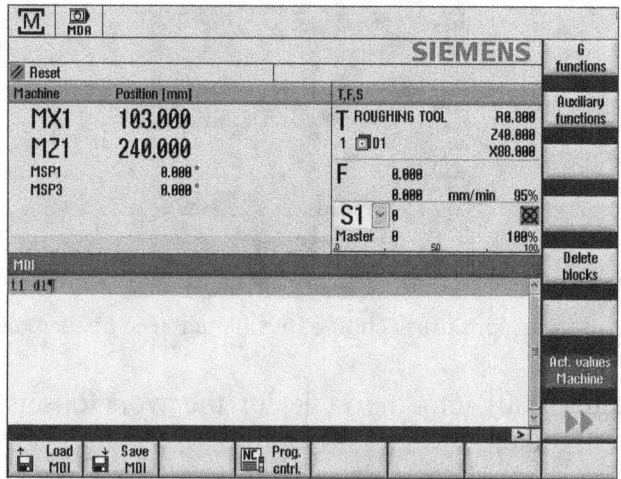

Fig. 59. Touch-off position of the part face in the machine coordinate system

Now return to MDA by pressing the relevant button.
Disable the button ACT. VALUES MACHINE.

Program G54 T1D1 and push CYCLE START (the functions for the newly set work offset and the tool offset values are being activated).

The current position of the tool on the Z-axis has become zero. As you can now see the part zero point is right on the front face of the touched workpiece.

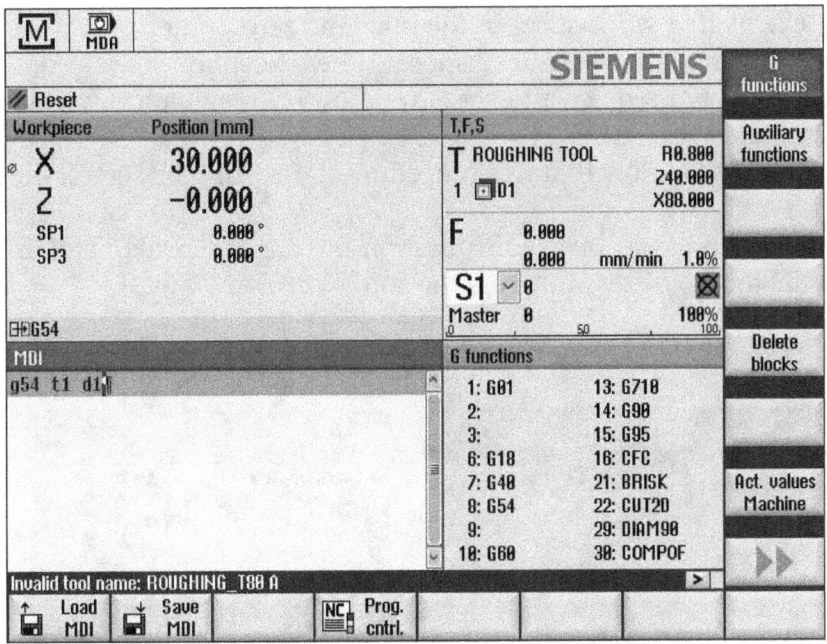

Fig. 60. Current position of the tool after activation of the workpiece coordinate system programmed in MDA

Attention: before going on press RESET to free the NC from the execution of the programmed block.

CNC – Basic Programming Course

6.3.3 Tool offset by touching the workpiece

In order to get the tool offset values in X and Z it is possible to measure the distance between the tool tip and the point moved by the NC manually, as shown in fig. 54.

Another procedure often used consists in touching the workpiece, entering the value and letting the machine calculate the value automatically.

On the X-axis, the value to enter in order to do the calculation is the diameter where the tool has touched the workpiece.

On the Z-axis, the value to enter is the distance between the touched face and the activated work offset or the machine zero point.

Having a simulator at our disposal, the execution of an operating procedure would require an excessive effort of imagination which might confuse rather than teach.

It is sufficient to know that this procedure starts from the screen shown in fig. 61.

In order to access it, push OFFSET on the control panel, TOOL LIST select the tool to offset and press the softkey MEAS. TOOL.

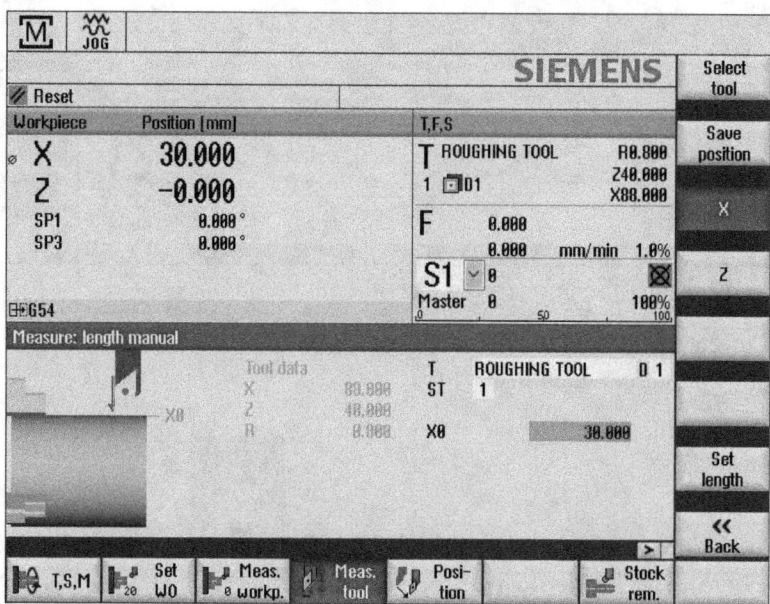

Fig. 61. Page for the automatic offset by touching the workpiece

7. Tool Call (2h)
(Theory: 1h, Practice: 1h)

7.1 Introduction
According to the logical programming sequence described in paragraph 5.2, after the setting of the part zero point, at the beginning of every operation, the tool and its corrector have to be selected.

7.2 T: tool call and function M6
By means of the address 'T', followed by the number of the position in which the tool is located, there is activation of the sequence of movements which permit the use of that tool for the tooling operation.

If the machine has a turret (as in this case), when reading the T instruction, the turret rotates to bring the tool onto the rotating axis of the workpiece. It is useful to remember that the rotating movement of the turret is an operation with a high collision risk. For this reason, it is good practice, before calling the tool, to move the turret into a safety position, which is normally programmed in machine coordinates.

Other types of lathes have a tool magazine located outside the working area. This has a higher number of available positions, but needs more time for the tool change. In this case the procedure for the positioning of the tool does not consist in the simple rotation of the turret, but rather in a sequence of actions such as the automatic positioning at the coordinates for the tool change, the pneumatic release of the attachment cone, the movement of the magazine to the tool deposit position and the remaining procedure of selecting the programmed tool.

This long sequence of operations is usually activated by the auxiliary function M6, which in some machines needs to be combined with the tool call instruction.

The turret of the analyzed machine has 20 positions.
The most common lathes in industry have turrets with six, eight, or twelve positions.
The number of tools which can be mounted in the machine is very important, because it shows the maximum number of tooling operations that the lathe is able to perform in one working cycle.
To go to the tool list page, press OFFSET on the control panel and then TOOL LIST from the horizontal softkeys.

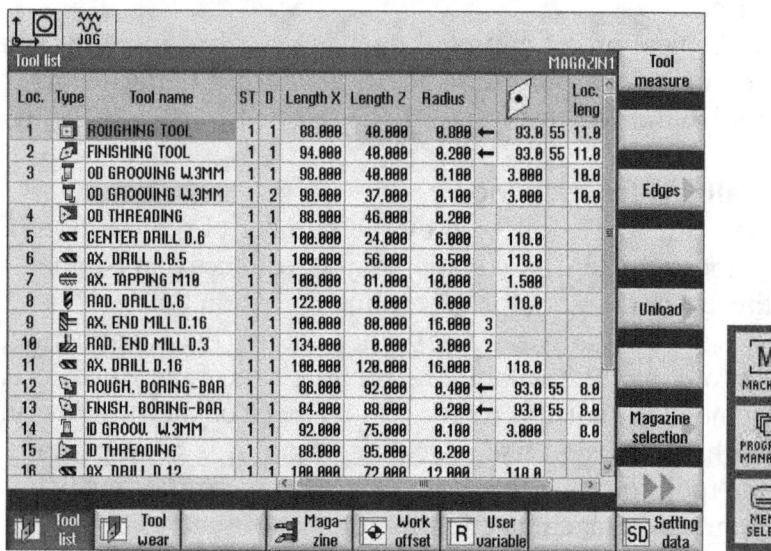

Fig. 62. Tool list page

The first column shows the physical location of the tool.
In the next practical exercise you will learn how to create and define, within the NC, the tools mounted to the turret.

7.3 D: tool offset values selection

For every tool it is necessary to define the table containing its offset values and the graphic description.
The address 'D' followed by the geometry table's progressive number activates the data contained therein.
These tables are also called CUTTING EDGES, because they set the position of the cutting edge controlled by the NC.

Various tables (up to 9) can be associated with every tool. This allows for the changing of the cutting edge used to describe the programmed profile. The simplest example to explain the use of a tool combined with a double offset is that of a groove.

Often, in order to manage the width of the groove, users prefer first to command the left hand cutting edge and then the right hand cutting edge. The left hand cutting edge is defined in table D1 and then the right hand cutting edge in table D2, where only the geometry value in Z is modified by a value corresponding to the insert width.

3		OD GROOVING W.3MM	1	1	98.000	40.000	0.100	3.000	10.0
		OD GROOVING W.3MM	1	2	98.000	37.000	0.100	3.000	10.0

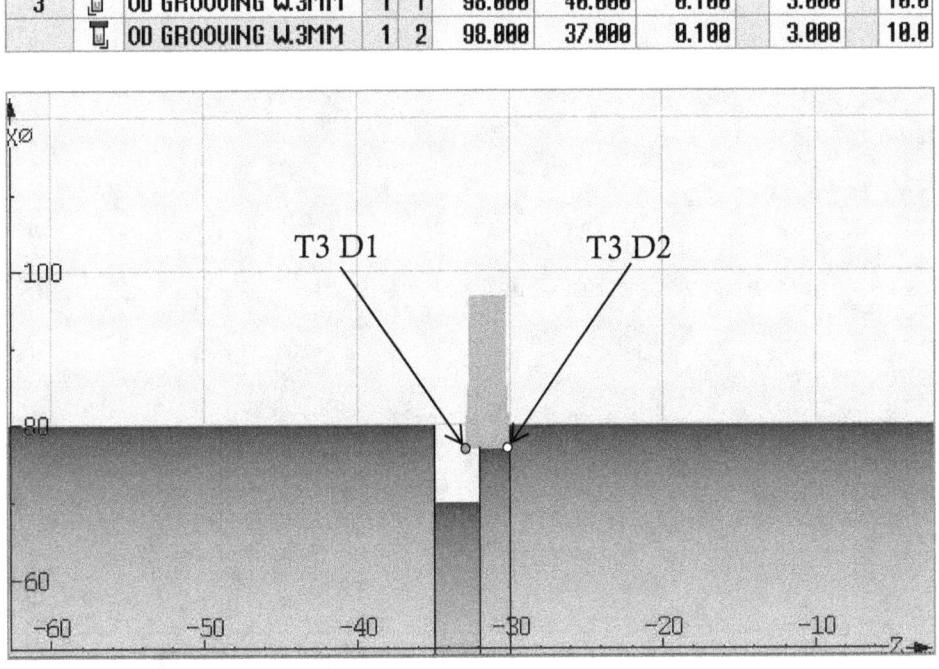

Fig. 63. Double corrector used for a 3 mm grooving tool

With the command D0 the tool offset values are deleted, with a return to command of the slide's characteristic point referring to the reference zero point active in that moment (part zero point or machine zero point). The following program (in the folder CHAP_07) creates a 5 mm groove, commanding first the left hand cutting edge (T3 D1) and then the right hand cutting edge (T3 D2).

```
; blank part dimensions:
; XA = 80 bar diameter
; ZA = 0 machining allowance on front face
; ZI = -200 length of finished part
; ZB = -150 extension from jaws

N10 WORKPIECE(,,,"CYLINDER",0,0,-200,-150,80)

N20 G18 G54 G90 ;G54 PART ZERO POINT SETTING
N30 G0 X400 Z500
N40 M8
N50 SETMS(1)

N60 T3 D1 ; GROOVING TOOL 3MM
; TABLE 1 DEFINES THE POSITION OF THE LEFT CUTTING
; EDGE
N70 G95 S1200 M4
N80 G0 Z-35
N90 G0 X82
N100 G1 X70 F0.12
N110 G0 X82

N120 D2 ; TABLE 2 DEFINES THE POSITION OF THE RIGHT CUTTING
; EDGE
N130 G0 Z-30
N140 G1 X70 F0.12
N150 G0 X82

N160 G0 X200
N170 G0 Z200
N180 M30
```

7.4 Correction of the tool wear

Every tool offset table is combined with a correction table, used by the operator to compensate the small variations on all axes in the machine due to the normal wear of the tool.

To go to the tool correction page press OFFSET on the control panel and then TOOL WEAR from the horizontal softkeys.

Loc.	Type	Tool name	ST	D	ΔLength X	ΔLength Z	ΔRadius	TC
1		ROUGHING TOOL	1	1	0.000	0.000	0.000	
2		FINISHING TOOL	1	1	0.000	0.000	0.000	
3		OD GROOVING W.3MM	1	1	0.000	0.000	0.000	
		OD GROOVING W.3MM	1	2	0.000	0.000	0.000	
4		OD THREADING	1	1	0.000	0.000	0.000	
5		CENTER DRILL D.6	1	1	0.000	0.000	0.000	
6		AX. DRILL D.8.5	1	1	0.000	0.000	0.000	
7		AX. TAPPING M10	1	1	0.000	0.000	0.000	
8		RAD. DRILL D.6	1	1	0.000	0.000	0.000	
9		AX. END MILL D.16	1	1	0.000	0.000	0.000	
10		RAD. END MILL D.3	1	1	0.000	0.000	0.000	
11		AX. DRILL D.16	1	1	0.000	0.000	0.000	
12		ROUGH. BORING-BAR	1	1	0.000	0.000	0.000	
13		FINISH. BORING-BAR	1	1	0.000	0.000	0.000	
14		ID GROOV. W.3MM	1	1	0.000	0.000	0.000	
15		ID THREADING	1	1	0.000	0.000	0.000	
16		AX. DRILL D.12	1	1	0.000	0.000	0.000	

Fig. 64. Tool correction page

It needs to be remembered that the potential use of a double corrector furthermore allows for correction of the tool when it is used for operations which require management independent of its wear, such as maintenance of a very tight tolerance during the finishing of multiple diameters.

7.5 Practical exercise

7.5.1 Creation of a tool

In all the exercises until this point we've used tools whose tooling data were imported from an external file according to the procedures laid down in paragraph 3.3.
It is now time to learn how to create new tools, how to set their offset values and the data used during the graphic simulation.

Press OFFSET, then press the TOOL LIST icon.
In order to display the icon which allows for the creation of a new tool it is important that no tool in the machine is selected.

Loc.	Type	Tool name	ST	D	Length X	Length Z	Radius		Loc. leng
1		ROUGHING TOOL	1	1	88.000	40.000	0.800 ←	93.0 55	11.0
2		FINISHING TOOL	1	1	94.000	40.000	0.200 ←	93.0 55	11.0
3		OD GROOVING W.3MM	1	1	98.000	40.000	0.100	3.000	10.0
		OD GROOVING W.3MM	1	2	98.000	37.000	0.100	3.000	10.0
4		OD THREADING	1	1	88.000	46.000	0.200		
5		CENTER DRILL D.6	1	1	100.000	24.000	6.000	118.0	
6		AX. DRILL D.8.5	1	1	100.000	56.000	8.500	118.0	
7		AX. TAPPING M10	1	1	100.000	81.000	10.000	1.500	
8		RAD. DRILL D.6	1	1	122.000	0.000	6.000	118.0	
9		AX. END MILL D.16	1	1	100.000	88.000	16.000	3	
10		RAD. END MILL D.3	1	1	134.000	0.000	3.000	2	
11		AX. DRILL D.16	1	1	100.000	120.000	16.000	118.0	
12		ROUGH. BORING-BAR	1	1	86.000	92.000	0.400 ←	93.0 55	8.0
13		FINISH. BORING-BAR	1	1	84.000	88.000	0.200 ←	93.0 55	8.0
14		ID GROOV. W.3MM	1	1	92.000	75.000	0.100	3.000	8.0
15		ID THREADING	1	1	88.000	95.000	0.200		
16		AX. DRILL D.12	1	1	100.000	72.000	12.000	118.0	

Fig. 65. Impossible to create a new tool when an already existing tool is selected

The icon with which to begin is NEW TOOL.

To make it appear, it is necessary to select an empty location in the turret or magazine.

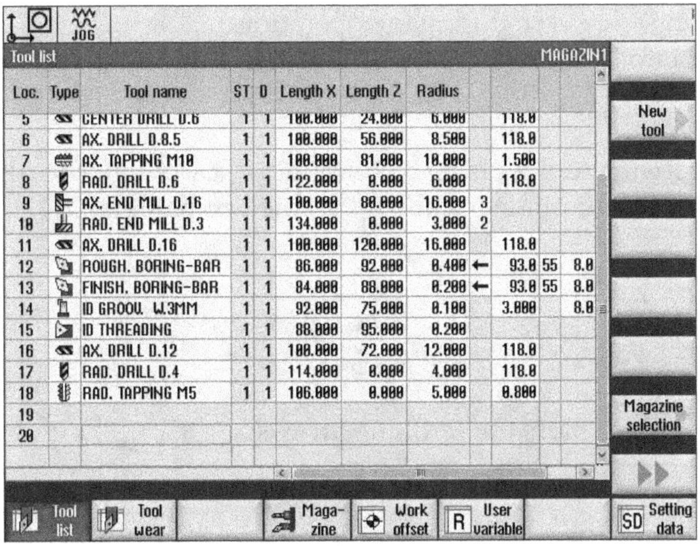

Fig. 66. Selection of an empty location for the creation of a new tool

Then press the vertical softkey NEW TOOL.

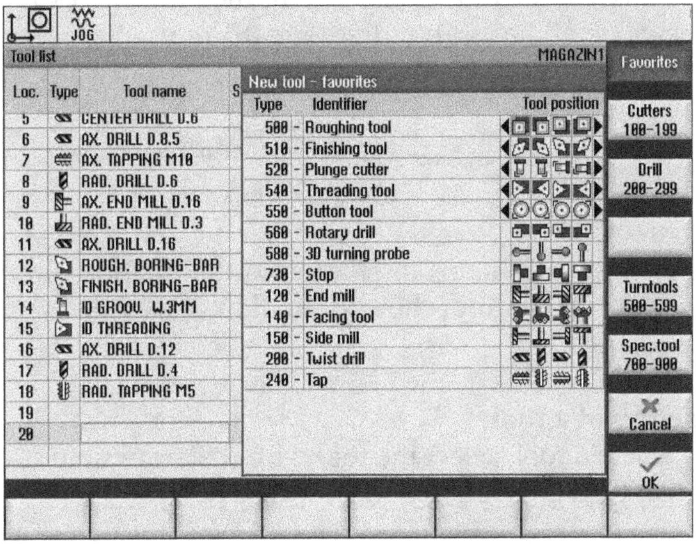

Fig. 67. Selection of the type of new tool to be created and position of the cutting edge

CNC – Basic Programming Course

The NC now proposes different tool types, mainly divided by type of operation performed (in the favorite list you find turning, cutting and drilling tools and tools for special operations), graphic aspect and position of the cutting edge used during the operation.
The position of the cutting edge is extremely important for the graphic simulation.

Select a grooving tool with zero point on the lower left side (the drawing refers to the programming scheme specified in paragraph 4.9) and press OK.

For more options press the scrolling arrows:

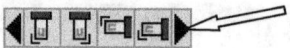

Rename the newly created tool as 'EXAMPLE' in order not to exchange it or overwrite it with another existing tool.
Always confirm the changes by pressing INPUT.
Now set the offset value in X (e.g. 80 mm), then the offset value in Z (e.g. 40 mm), the value of the insert radius (e.g. 0.1 mm), the insert width (e.g. 3 mm) and its length (e.g. 10 mm).

20	EXAMPLE	1	1	80.000	40.000	0.100	3.000	10.0

Fig. 68. Creation of a new tool

Leave out the last three items in relation to the rotation direction of the spindle and the activation of the cooling liquid, as they are not used in ISO programming.

7.5.2 Deletion of a tool
In order to delete a tool, select the tool with the arrows, press the vertical softkey DELETE TOOL and confirm with OK.

7.5.3 Creation of a second tool corrector

As already seen in paragraph 7.3, a tool can be associated with various cutting edges. Now, define the secondary edge of a grooving tool according to the following procedure:

- Press OFFSET

- Make sure you are in the TOOL LIST

- With the arrows, highlight the tool you want to combine with another cutting edge

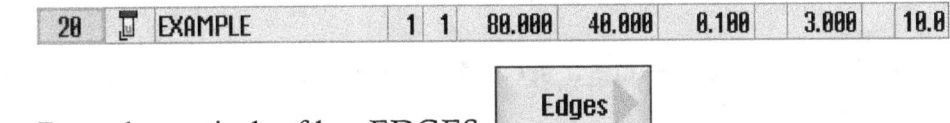

- Press the vertical softkey EDGES

- Press the vertical softkey NEW CUTTING EDGE

- A new window opens with the same tool offset values, though identified by the number 2 in column 'D'.

- Now select the box for the cutting edge, press SELECT and select the one down on the right.

- Now, vary the offset values according to the position of the new cutting edge. In this case, the second edge is located in the same position in X and changes in Z by a value which corresponds to the width of the insert. Enter the new value 37 (derived from: 40 – 3) in the Z column.

7.5.4 Deletion of a second tool corrector
In order to delete a second corrector formerly created:

- Select the edge you want to delete with the cursor.

20	EXAMPLE	1	1	80.000	40.000	0.100	3.000	10.0
	EXAMPLE	1	2	80.000	37.000	0.100	3.000	10.0

- Press the vertical softkey EDGES

- Then press the softkey DELETE CUTTING EDGE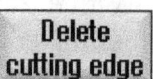

Make sure not to delete a second corrector by pressing the softkey DELETE TOOL.

7.5.5 Mounting and removal of the tools in the turret

The lines from 1 to 20 represent the available locations on the turret. The following lines show all the tools created but not mounted to the turret, as if those were stored in an external magazine.

In order to remove a tool from the first 20 positions, select the tool you want to remove with the arrows and press the vertical softkey UNLOAD. In the opposite direction, to mount a tool archived in the following 20 locations, select the tool and press the vertical softkey LOAD. Automatically, a free location is proposed; change it if necessary and confirm with OK.

7.5.6 Saving of tooling data (with license only)

Paragraph 3.3. contains the procedure to import tool data from an external file. Now we will see how to save these data in the folder containing the main program that has used them.

Press PROGRAM MANAGER on the control panel.
Select the folder in which you want to save the tooling data.
Press ARCHIVE in the list of vertical softkeys.
Then press SAVE SETUP DATA.

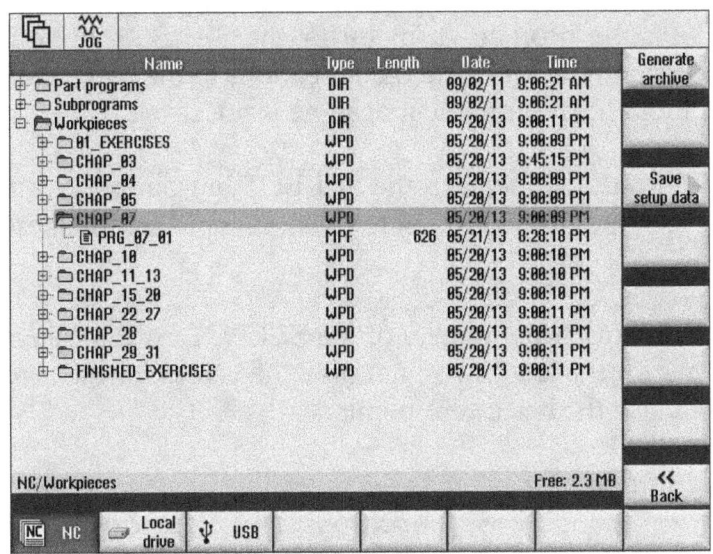
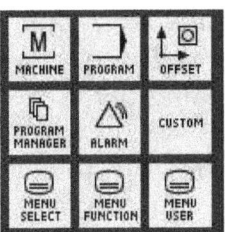

Fig. 69. Saving of tooling data

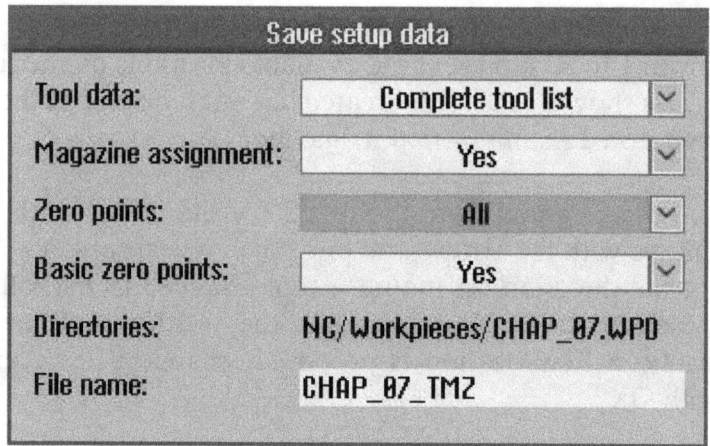

Fig. 70. Window for the saving of tooling data

Tool data: with the drop-down menu or the SELECT button placed at the center of the arrows, select *Complete tool list* to save the complete list of the tools in the machine. If you select *No* this means that you do not want to save the tool data but only the zero points (from G54 to G57) and the basic zero points (BASE).
Move with the arrows onto the next items.

Magazine locations: with the drop-down menu or the SELECT button, select *Yes*. This option saves the tools and their locations in the magazine. If you select *No*, the tool's location in the magazine is not saved.

Zero points: with the drop-down menu or the SELECT button, select *All*. This option saves the zero points (from G54 to G57). If you select *No* these data will not be saved.

Basic zero points: with the drop-down menu or the SELECT button, select *Yes*. This option allows for the loading of, not only the values for the zero points of the axes, but also the basic zero points.

Press OK to save the current data.

8. Spindle Activation (2h)
(Theory: 1.5h, Practice: 0.5h)

8.1 Introduction

The number of revolutions of the spindle and the diametrical position of the tool are the data which define one of the most important cutting parameters of a tooling operation: the cutting speed.

The cutting speed is the speed with which the chip moves on the cutting edge of the tool.

The general concept of speed expresses the distance traveled in a certain unit of time.

In one revolution, the distance traveled by a point rotating on a certain diameter corresponds to the circumference that the diameter itself defines, calculated according to the formula shown in the following figure.

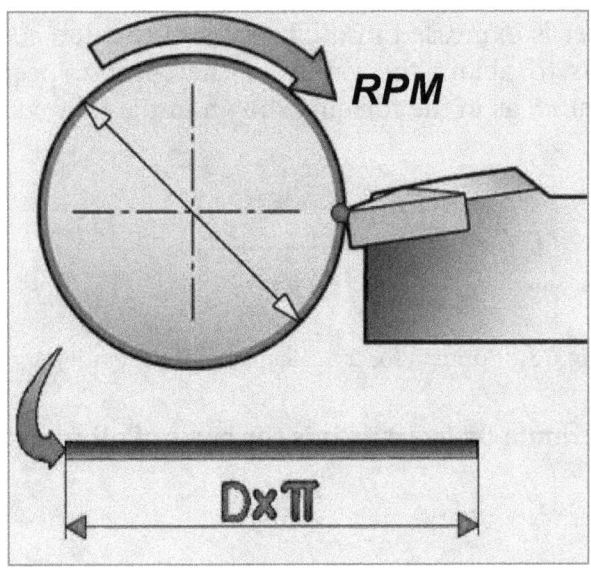

Fig. 71. Distance traveled by the tool in one revolution

By multiplying the circumference by the number of revolutions per minute of the spindle the total distance traveled by the tool in one minute is calculated.

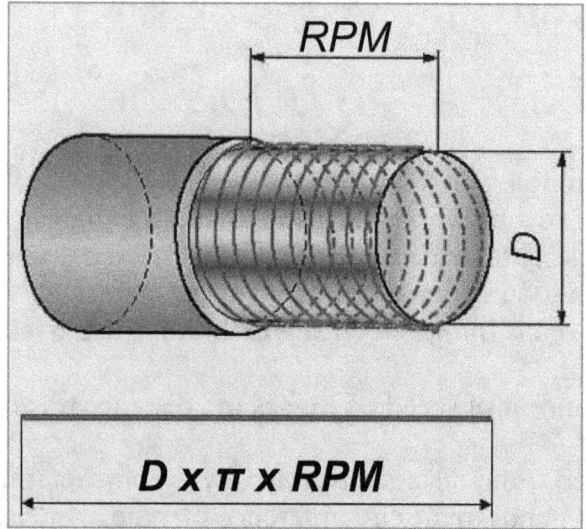

Fig. 72. Distance traveled by the tool in one minute with rotating workpiece

As the diameter is expressed in millimeters, this value is divided by one thousand so as to obtain the value for the cutting speed expressed in meters per minute, as in the formula shown in the following figure.

$$C_s = \frac{D \times \pi \times RPM}{1000} \quad \frac{m}{min}$$

Fig. 73. Formula for the calculation of the cutting speed

Learn these formula by heart as it is the basis of all turning technology.

8.2 SETMS: setting of the master spindle

In the program, before setting the rotation and feed in millimeters per revolution functions, it is necessary to define the spindle to which these refer.

The SETMS (Set Master Spindle) function, followed by the name of the spindle, defines the master or reference spindle.

On the page showing the current coordinates, other than the X- and Z-axis, the angular spindle positions are also shown: SP1 and SP3.

SP1 is the name of spindle '1' (workpiece carrying spindle).
SP3 is the name of spindle '3' (driven tools spindle).

SP2 would be the name of any counterspindle present in the machine.

Attention: Not all lathes use the same terms. Therefore refer to the manufacturer's manual.

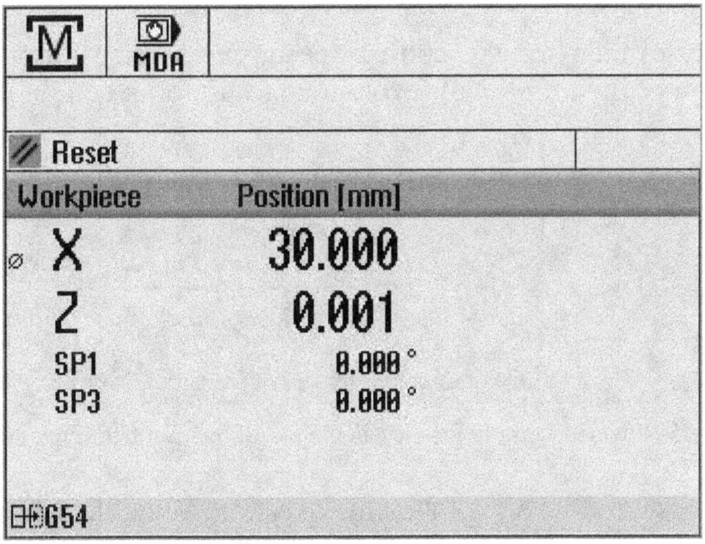

Fig. 74. Name of the spindles shown on the "current position" page

At the beginning of the programs that have been used so far, the SETMS(1) function is always programmed. This function sets the spindle holding the workpiece as master spindle, i.e. it says that all the rotation and feed functions (in mm/rev) programmed after its activation refer to spindle no. '1'.

8.3 G97: Spindle rotation with constant number of revolutions

The function G97 sets the spindle rotation to a constant number of revolutions. The number of revolutions is programmed using the address 'S'.

Example: **G97 S1000**.

The rotation direction of the spindle will be explained in paragraph 8.6.

The cutting speed depends on:
- material to be worked on (aluminum, steel, titanium, etc.),
- material of the tool used (HSS, sintered carbides)
- type of tooling operation (roughing, finishing, cutting, drilling)
- cutting conditions (workpiece extending much from spindle)

Its value, suggested by the tools' manufacturers or obtained from operator experience, is the only certain data item from which to start the calculation of the number of revolutions.

From the formula for the cutting speed, the inverse formula for the calculation of the number of revolutions to use on a certain diameter is obtained.

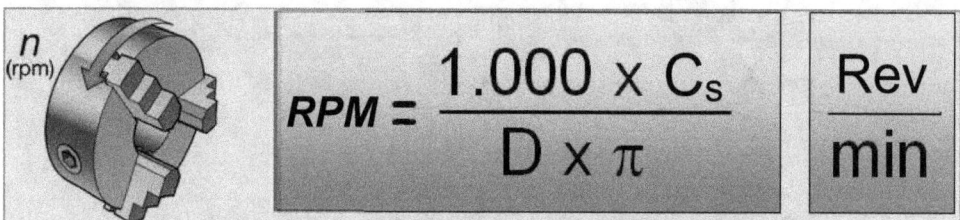

Fig. 75. Inverse formula for the calculation of the number of revolutions

We are now faced with the following question: What if, as happens with the facing or cutting of a workpiece, the diameter changes during the tooling operation?

The answer is that the cutting speed will change accordingly. With a constant number of revolutions, the cutting speed will be lesser on diameters smaller and greater on diameters larger than the one calculated, following the trend indicated by the graph in the following figure.

On the x-coordinate (the horizontal axis) we have the diameter value, on the y-coordinate (the vertical axis) we have the cutting speed value.

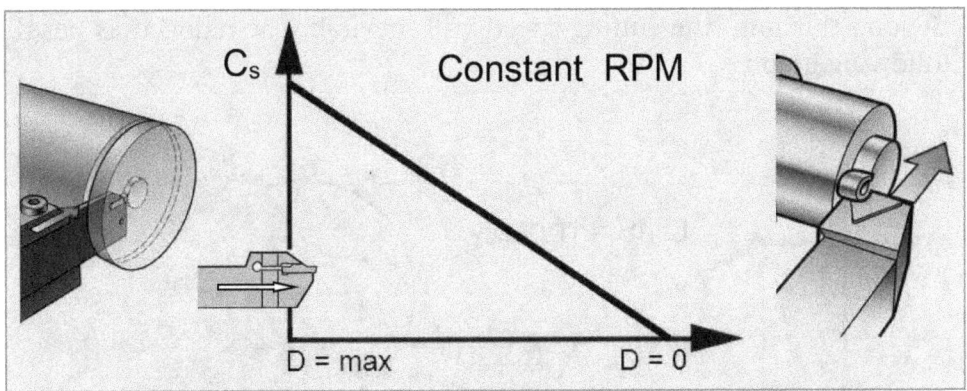

Fig. 76. Graph for the evolution of cutting speed at variation of working diameter, at constant number of revolutions

8.4 G96: setting of constant cutting speed

In order to maintain a constant cutting speed, a different function belonging to the same group as G97 needs to be used.

Indeed, calculation of the revolutions necessary to maintain constant cutting speed based on the diameter on which the tool is working is delegated to the machine by the function G96, followed by the address 'S' and by the value for the cutting speed (example: **G96 S120**).

After the programming of G96, the number of revolutions automatically adapts to the diameter of the workpiece: the higher the diameter is, the lower the number of revolutions is; the lower the diameter is, the higher the number of revolutions is.

If we now take the inverse formula for the calculation of the number of revolutions and if we perform a hypothetical facing operation up to a diameter of 2mm, we obtain as a result at constant cutting speed of 100 meters per minute a value for the spindle rotation of 15923 revolutions per minute.

If one were to consider the diameter to be zero, this value would even have to be infinite.

It is clear that the operator must be able to set the maximum number of revolutions achievable by the spindle.

This requirement leads us into the next paragraph.

8.5 LIMS=: limitation of the maximum number of revolutions

The command 'LIMS' sets a maximum limit for the number of revolutions of the master spindle (e.g. **LIMS=4000**).
Beyond this limit the cutting speed will inevitably be reduced as per the following graph.

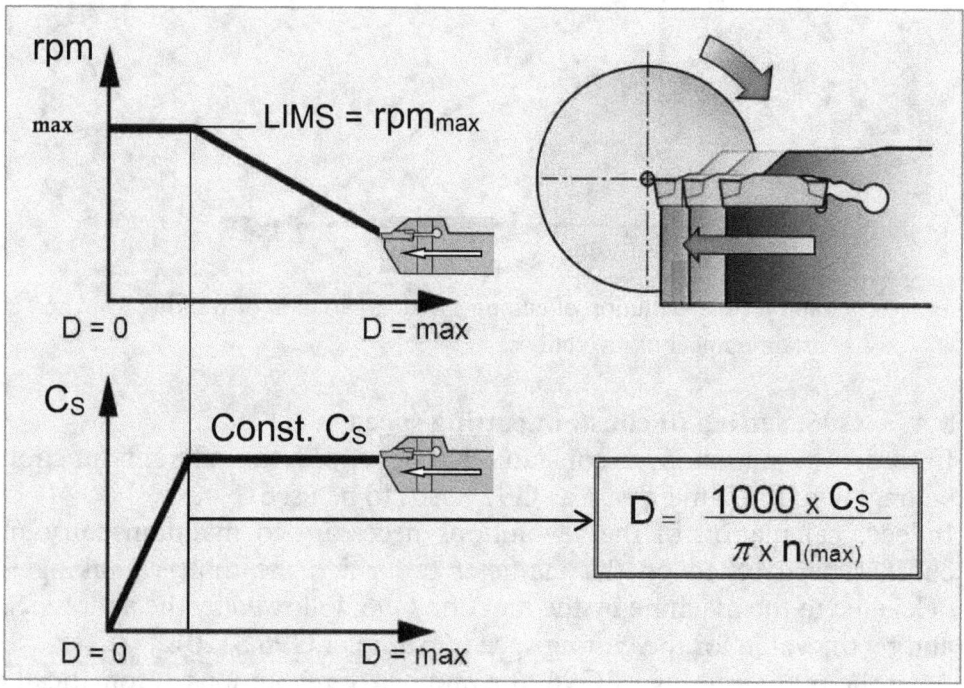

Fig. 77. Graph for the cutting speed trend beyond the number of revolutions threshold

Starting with the biggest diameter and approaching the smallest one, the NC holds the cutting speed constant and increases the number of revolutions up to the threshold set by the LIMS function.
From that point on, the cutting speed begins to drop starting from a diameter D (calculated by means of the formula shown in the figure) until it becomes zero (Cs=0) at the rotation center of the workpiece.

8.6 M3, M4, M5: setting of the rotation direction

The auxiliary functions M3 and M4 set the rotation direction of the spindle, in clockwise or counterclockwise direction.
Conventionally, the direction is defined from the point of view of an observer standing behind the spindle.
According to the structural features of the machine, the manufacturer suggests the ideal rotations direction.
What varies when turning in one direction rather than the other is the choice of tool type; the clockwise rotation (M3) requires the use of right hand drills, while the counterclockwise rotation (M4) requires the use of left hand drills.
The rotation direction is therefore also influenced by the characteristics of the tools available in the magazine.
The configuration of the machine under examination suggests a counterclockwise rotation (M4), because this allows the cutting stress to be discharged through the machine base and the insert to point towards the operator (simplifying the checking and replacement thereof).
The auxiliary function M5 stops the rotation of the spindle.

8.7 Instructions to a spindle which is not the master spindle

As already seen in paragraph 8.2 the SETMS function sets the master spindle.
This is the spindle to which refer instructions for speed ('S'), rotation direction ('M'), limitation of the number of revolutions ('LIMS=') and of angular orientation ('SPOS=').
If you want to use these functions in order to control a spindle which is not defined as the master spindle, or to display clearly the name of the spindle next to the function, you can use the following programming syntax:

```
G96 S1=120         ; S120 refers to the spindle named 1
G97 S1=2200        ; S2200 refers to the spindle named 1
G97 S3=1600       ; S1600 refers to the spindle named 3
M1=3               ; M3 refers to the spindle named 1
M1=4               ; M4 refers to the spindle named 1
M3=3               ; M3 refers to the spindle named 3
LIMS[1]=4000       ; LIMS=4000 refers to the spindle named 1
```

8.8 Choice of the functions G97, G96 and LIMS

The constant number of revolutions (G97) is programmed in the following cases:
- the diameter to be worked on does not change (cylindrical turning),
- fluctuations in the number of revolutions are not desired (execution of a thread in multiple passes),
- the working diameter is zero and therefore the number of revolutions would be incalculable (as for on-axis-drillings, where the number of revolutions is to be calculated according to the cutting speed and the diameter of the drill).

Constant speed is programmed (G96) when:
- the working diameter changes significantly (facing, parting-off, profiling of the workpiece).

The limit of the maximum number of revolutions is set (LIMS):
- every time that G96 is used in the program.

8.9 SPOS=: programming of the angular orientation

The option offered by a lathe to mount driven tools is always combined with its capability of orientating the spindle at an angle. This allows the execution of radial on-axis drilling and milling operations, or out-of-axis longitudinal (frontal) holes.

With SPOS it is possible to place the spindles in certain angular positions. **The simple angular orientation is not considered to be an axis on its own as it is not able to interpolate with any other axis in the machine (see paragraph 4.1).**

The value that follows the SPOS function expresses the angle for the positioning of the spindle with reference to its zero point; this is expressed by a value between 0 and 360 degrees.

The programming is carried out as follows:

```
SPOS=0      ; angular orientation at zero degrees of the
              spindle defined as master spindle by means of the
              SETMS function

SPOS[1]=0   ; angular orientation at zero degrees of the
              spindle defined as no. '1' even if not defined as
              master spindle
```

8.10 Practical exercise

8.10.1 Calculation exercises

Based on the data referring to working diameter, number of revolutions and cutting speed, calculate and write down the missing data item in the relevant field.

Working diameter (mm)	Number of revolutions (r/min)	Cutting speed (m/min)
50	764	120
62	…………………	140
19	…………………	85
5	…………………	100
55	1200	…………………
8	1200	…………………
62	650	…………………
…………………	4500	100
…………………	2000	40
…………………	2000	220

Fig. 78. Exercise for the calculation of the cutting speed, the number of revolutions and the diameter from which the cutting speed begins to decrease

8.10.2 Creation of a new main program

The following paragraph is not bound to the topics covered in this chapter, as it describes the procedure for the creation of a new program.

Press PROGRAM MANAGER on the control panel.

With the arrows or with your mouse, select the system folder PART PROGRAMS.

This folder is designed only to contain main programs with .MPF (Main Program Files) extension.

Press NEW.

Press the vertical softkey PROGRAM GUIDE G CODE in order to create a program developed in ISO language, and not by means of the conversational software Siemens ShopTurn.

Enter the name of the new program (e.g. WORKPIECE_1).

Confirm with OK.

The newly created empty program opens automatically.

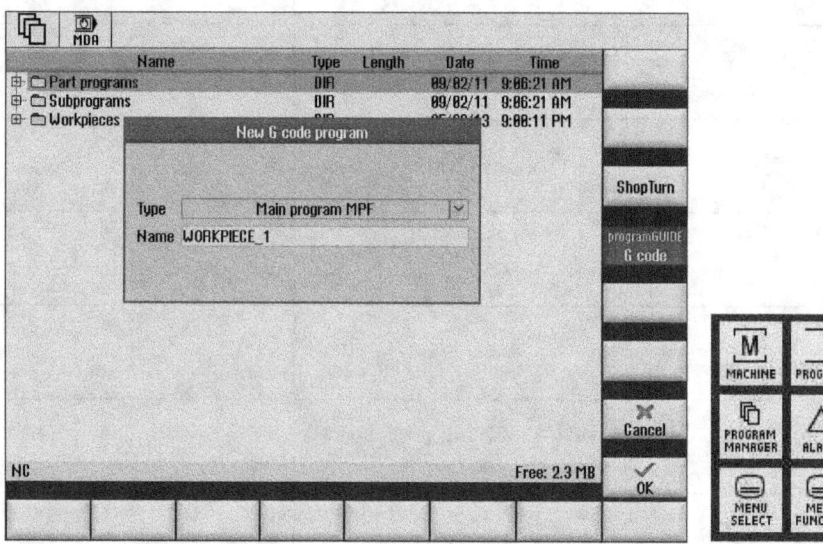

Fig. 79. Creation of a new program

Close the program by pressing NEXT 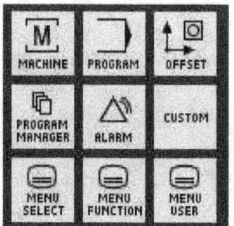 and then EXIT.

We are not going to use the PART PROGRAMS folder as it does not allow us to organize the programs into further subfolders.

8.10.3 Creation of a new subprogram

Press PROGRAM MANAGER on the control panel.
With the arrows or with your mouse, select the system folder SUBPROGRAMS.
This folder is designed only to contain subprograms called by main programs. Their extension is .SPF (Sub Program Files).
Press NEW.
Press the vertical softkey PROGRAM GUIDE G CODE in order to create a subprogram in ISO language.
Enter the name of the new subprogram (e.g. SUB_1).
Confirm with OK.
The newly created empty subprogram opens automatically.
Close it by following the procedure laid down in the previous paragraph.

8.10.4 Creation of a new folder

Press PROGRAM MANAGER on the control panel.
With the arrows or with your mouse, select the system folder WORKPIECES.
This folder is designed only to contain subfolders which can contain main programs and subprograms.
The extension of the folders is .WPD (Work Piece Directory).
Press NEW.
Enter the name of the new folder (e.g. FOLDER_1).
Confirm with OK.
The new folder will be listed together with the other folders in alphabetical order. A window will automatically open asking you if you want to create a main program with the same name in the folder.
You can choose to change the name, and, by means of the drop-down menu, select if you want it to be a main program (.MPF) or a subprogram (.SPF).
Maintain the selection MAIN PROGRAM MPF and press OK once again. Close the newly created program.
In order to create several .MPF or .SPF programs in the same folder, select the folder and proceed as described in paragraphs 8.10.2 and 8.10.3.

8.10.5 Spindle angular orientation exercises review
Repeat the exercises contained in paragraphs 4.10.1 and 4.10.2 and analyze the programming syntax for the angular orientation of the spindle.

9. Setting of the Feed rate (1h)
(Theory: 0.5h, Practice: 0.5h)

9.1 Introduction
The 'F' address expresses the value for the feed rate used during the tooling operations. Based on the instruction entered in the same block or on the active modal instruction, its value can be expressed in millimeters per revolution (G95) or in millimeters per minute (G94).

9.2 G95: feed rate expressed in mm/rev
The feed rate of a tool in a lathe is normally expressed in millimeters per revolution.
In this case the translation speed of the tool varies according to the number of revolutions of the spindle.

Attention: G95 also sets the fixed number of revolutions. This is why it often replaces function G97, as programmed in the following example block.

```
G95 S1800 M4      ; setting of the constant revolutions and
                  ; of the feed rate in millimeters per rev.
```

9.3 G94: speed in mm/min
G94 sets the feed rate of the tool as translation speed of the slide expressed in millimeters per minute.
This value remains totally unrelated to the number of revolutions of the spindle.
Multiplying the feed rate value expressed in millimeters per revolution (F_{rev}) by the number of revolutions per minute (RPM), the equivalent feed value expressed in millimeters per minute (F_v) is obtained.

$$F_v = F_{rev} * RPM$$

9.4 Calculation of the execution time for one pass

This paragraph analyzes the calculation method for the time the tool needs to execute one pass.

Assuming a cut lungth of 50 millimeters (L), a feed rate of 0.2 millimeters per revolution (F_r) and a spindle rotation of 1400 revolutions per minute (RPM), proceed as follows:

- Calculate the distance that the tool travels in one minute by using the formula shown in paragraph 9.3.

$$F_v = (F_r * RPM) = 0.2 * 1400 = 280 \text{ mm/min}$$

- From here you can calculate the speed expressed in millimeters per second.

$$v_s = F_v / 60 = 280 / 60 = 4.66 \text{ mm/sec}$$

- Dividing the length to be traveled by the execution speed of the pass, the time necessary to complete the pass is obtained.

$$t = L / v_s = 50 / 4.66 = 10.72 \text{ sec}$$

Combining these formulas you obtain:

$$t = L * 60 / (F_r * n)$$

This formula is useful in calculating in advance the time needed for the production of one piece.

9.5 Practical exercise

9.5.1 Calculation exercises

Based on the data referring to: length, feed rate and number of revolutions, calculate the time the tool needs to execute the pass.

Cut length (mm)	Feed rate (mm/rev)	Number of revolutions (rpm)	Time necessary (seconds)
60	0.3	840	....
60	0.12	1100	....
24	0.1	1260	....
18	0.06	780	....
22	0.14	1530	....
80	0.18	2100	....
66	0.05	1400	....
43	0.25	600	....

Fig. 80. Calculation exercises for the time needed for the tool to execute a pass

9.5.2 Saving of folders and programs

This paragraph explains how to save a folder or a program from the NC's memory on an external memory (usually with USB connection).

Before going on, connect a USB memory to the computer.

Press PROGRAM MANAGER on the control panel.

With the arrows or the mouse, select the program or the folder you want to save (e.g. the program PRG_03_01).

Press COPY from the list of vertical softkeys.
Then press USB in order to select the target device on which to save the data.
Press PASTE in order to start the copying process of the data.

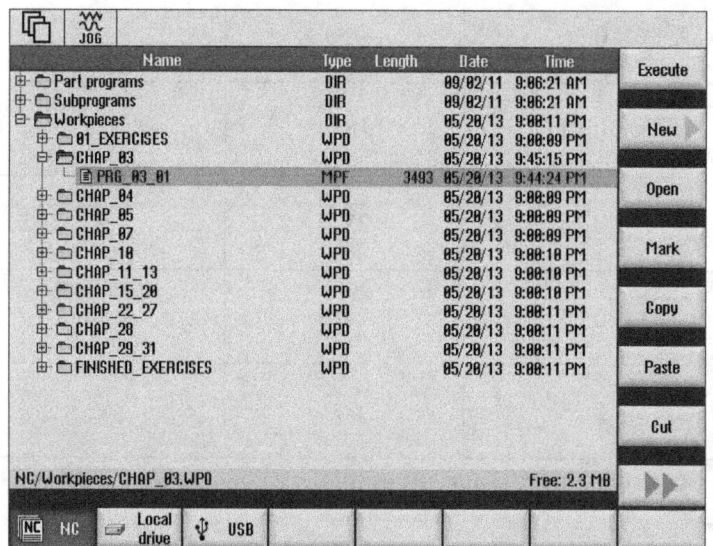
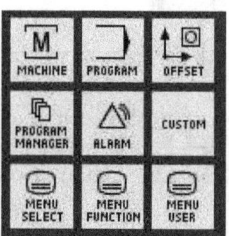

Fig. 81. Saving of folders and programs in an external memory

Press NC again in order to return to the NC's memory.

10. Absolute and Incremental Coordinates (1h)
(Theory: 0.5h, Practice: 0.5h)

10.1 G90: absolute programming

The function G90 sets the absolute coordinate system; this allows all coordinates expressed in the program to be referred to a single point, which can be the part zero point or the machine zero point. G90 is already enabled at the machine start up.

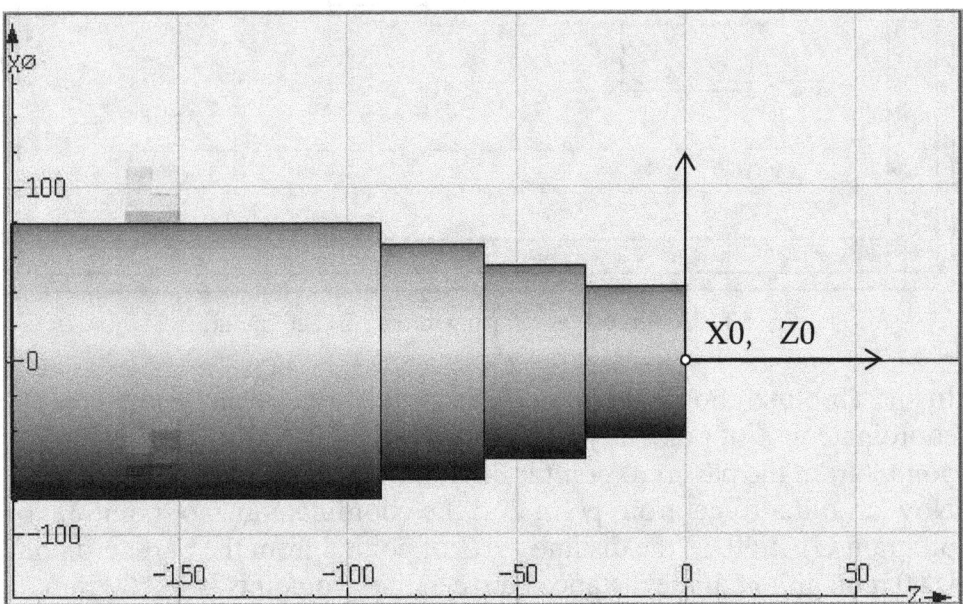

Fig. 82. Origin of the axes in the absolute coordinate system referring to the part zero

In order to support the operator in the programming of the profile, the designer often refers all the values to the front face of the workpiece.

The following drawing shows a workpiece with three steps of length 30 mm.

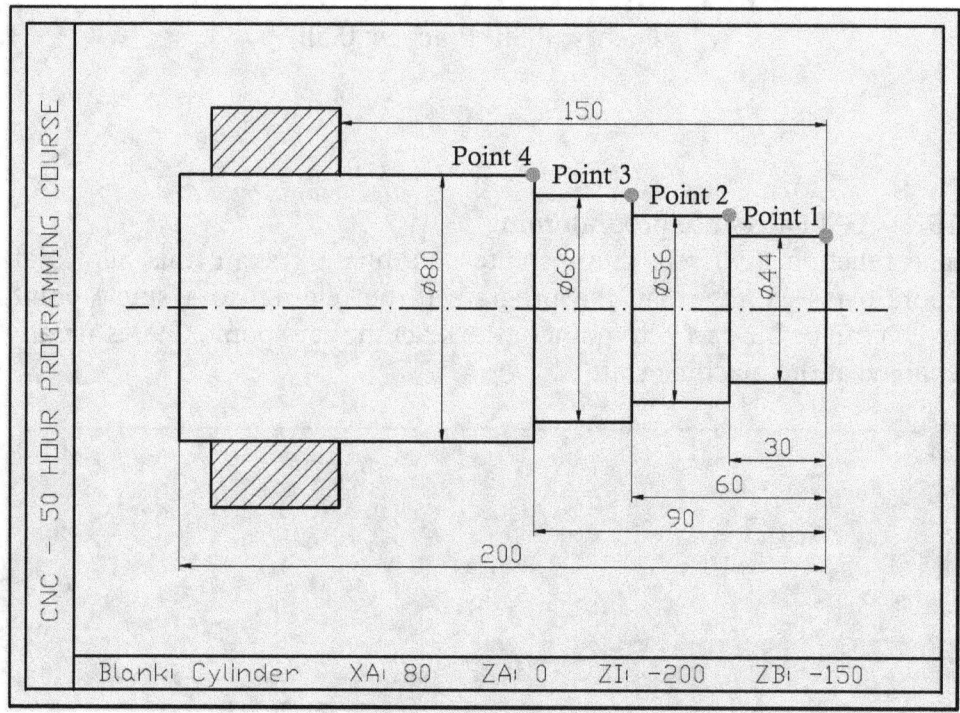

Fig. 83. Design values referring to the part zero point.

In the absolute coordinate system, in order to get from point 1 to the coordinate in Z of point 2, you program G1 Z-30, as the distance in Z of point 2 from the part zero point is 30 mm.

Now, in order to get from point 2 to the coordinate in Z of point 3, you program G1 Z-60, as the distance in Z of point 3 from the part zero point is 60 mm, although the distance between the two points is still 30 mm.

Now, in order to get from point 3 to the coordinate in Z of point 4, you program G1 Z-90, as the distance in Z of point 4 from the part zero point is 90 mm.

10.2 G91: incremental programming

The function G91 sets the incremental coordinate system.
When G91 is enabled, all coordinates refer to the current position of the tool.
The location of the tool becomes the zero point to which the subsequent movement refers.

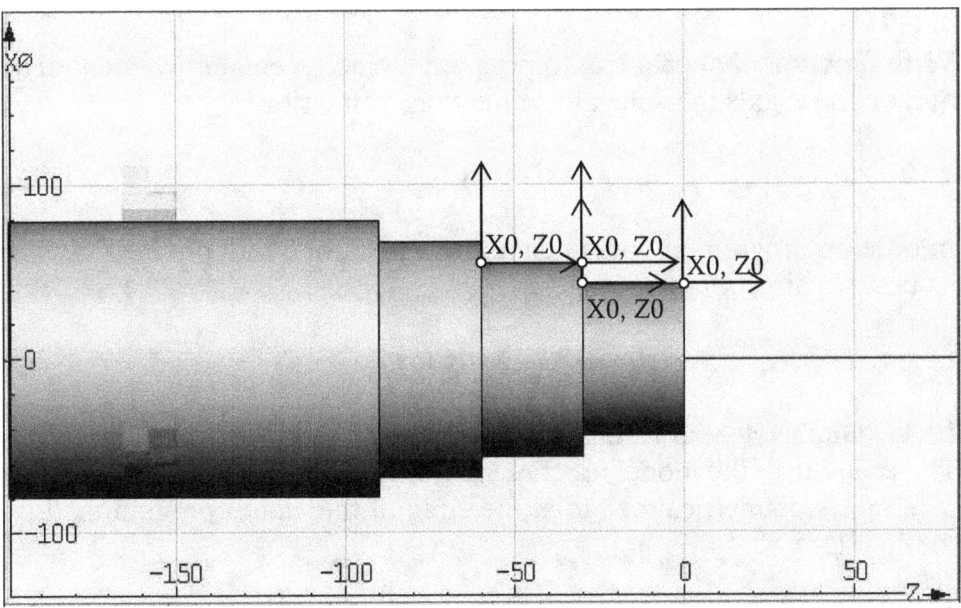

Fig. 84. Origin of the axes in the incremental coordinate system

The incremental coordinate system is never used for the definition of an entire profile. It is nonetheless useful (without ever being necessary) in defining the width of a groove or in programming a hole with chip removal.

Later we will see that incremental values are necessary when we want to repeat program parts, shifting their starting point (various holes executed with constant incrementation in Z), though in this case you can also use the mixed programming without necessarily enabling the function G91, as described in the following paragraph.

10.3 Mixed programming

By means of the function G90 it is possible to program incremental movements.

The possibility to enter values expressed both in absolute and in incremental coordinates in the same block is what gives this programming type its name.

The following programming syntax in most of the cases replaces the use of G91.

With function G90 enabled, to program an incremental movement of 60 mm on the Z-axis in its negative direction, you write:

$$G1\ Z=IC(-60)$$

In order to program an incremental movement of 5 mm on the X-axis in its positive direction, you write:

$$G1\ X=IC(5)$$

10.4 Diametrical or radial meaning of the values associated with X

The enabling of the modal functions DIAMON, DIAM90 and DIAMOF determines diametrical or radial meaning of the values programmed on the X-axis.

DIAMON attributes diametrical meaning to all values associated with X.

DIAM90 sets diametrical meaning in the absolute coordinate system (G90), while it attributes radial meaning when the value is expressed in the incremental coordinate system, both when it is enabled by function G91 or by means of the syntax X=IC(...).

DIAMOF attributes radial meaning to all values associated with X.

Attention: in the lathe that we are using, the active modal function at machine start is DIAM90.

10.5 Practical exercise

10.5.1 Analysis of a program in absolute coordinates

Open the program PRG_10_01 in the folder CHAP_10, start the graphic simulation and enable the single block mode. This program creates the part described in the drawing in fig. 83. Analyze the value of the programmed coordinates in every single block and monitor the tool's movement.

```
; blank part dimensions:
; XA = 80 bar diameter
; ZA = 0 machining allowance on front face
; ZI = -200 length of finished part
; ZB = -150 extension from jaws
N10 WORKPIECE(,,,"CYLINDER",192,0,-200,-150,80)

N20 G18 G54 G90 ;G54 PART ZERO POINT SETTING
; G90 ABSOLUTE COORDINATE SYSTEM
N30 G0 X400 Z500
N40 M8 ; COOLANT ACTIVATION
N50 SETMS(1) ; SETTING OF MASTER SPINDLE

N60 T1 D1 ; ROUGHING TOOL
N70 G95 S1800 M4 F0.2 ; SETTING OF NUMBER OF REVOLUTIONS AND
FEED RATE IN MM/REV

N80 G0 X68 Z2
N90 G1 Z-90
N100 G1 X82
N110 G0 Z2

N120 G0 X56
N130 G1 Z-60
N140 G1 X70
N150 G0 Z2

N160 G0 X44
N170 G1 Z-30
N180 G1 X58
N190 G0 Z2

N200 G0 X200
N210 G0 Z200
N220 M30
```

10.5.2 Analysis of a program in incremental coordinates

Open the program PRG_10_02 in the folder CHAP_10 and start the graphic simulation in single block mode. The same workpiece created on the previous page is now programmed in incremental coordinates using the mixed programming. Analyze the value of the programmed coordinates in every single block and monitor the tool's movement. Note how abuse of the incremental programming makes the program difficult to understand.

```
...
N20 G18 G54 G90 ;G54 PART ZERO POINT SETTING
; G90 ABSOLUTE COORDINATE SYSTEM
N30 G0 X400 Z500
N40 M8 ; COOLANT ACTIVATION
N50 SETMS(1) ; SETTING OF MASTER SPINDLE

N60 T1 D1 ; ROUGHING TOOL
N70 G95 S1800 M4 F0.2 ; SETTING OF NUMBER OF REVOLUTIONS AND
FEED RATE IN MM/REV

N80 G0 X68 Z2 ; POSITIONING IN ABSOLUTE COORDINATES
N90 DIAMON ; VALUE OF INCREMENTAL COORDINATES IN X WITH
DIAMETRAL MEANING

N100 G1 Z=IC(-92) ; ABSOLUTE COORD. Z-90
N110 G1 X=IC(14) ; ABSOLUTE COORD. X82
N120 G0 Z=IC(92) ; ABSOLUTE COORD. Z2

N130 G0 X=IC(-26) ; ABSOLUTE COORD. X56
N140 G1 Z=IC(-62) ; ABSOLUTE COORD. Z-60
N150 G1 X=IC(14) ; ABSOLUTE COORD. X70
N160 G0 Z=IC(62) ; ABSOLUTE COORD. Z2

N170 G0 X=IC(-26) ; ABSOLUTE COORD. X44
N180 G1 Z=IC(-32) ; ABSOLUTE COORD. Z-30
N190 G1 X=IC(14) ; ABSOLUTE COORD. X58
N200 G0 Z=IC(32) ; ABSOLUTE COORD. Z2

N210 G0 X200
N220 G0 Z200
N230 M30
```

11. Basic Functions to Define the Profile (3h)
(Theory: 1h, Practice: 2h)

11.1 G0: rapid movement

As already seen in the programs used so far, before the tooling operations, one or more blocks for approaching the tool to the workpiece are always programmed by means of function G0.

G0 sets the rapid movement of the slide or the tool to the programmed point. The speed of the rapid movements is pre-set by the manufacturer and depends on the machine characteristics. For a lathe such as the one examined here, a maximum rapid movement speed of 30,000 mm/minute (30 meters per minute) is already ideal.

The modal function Siemens RTLION is active upon machine start and sets the linear trajectory of the rapid path.

The command RTLIOF overwrites this and sets the reaching of the arrival point without linear interpolation, thereby achieving a faster positioning speed but also increasing the risk of a collision.

When you enter this function into the program, be careful not to write GO (letter O) instead of G0 (number zero).

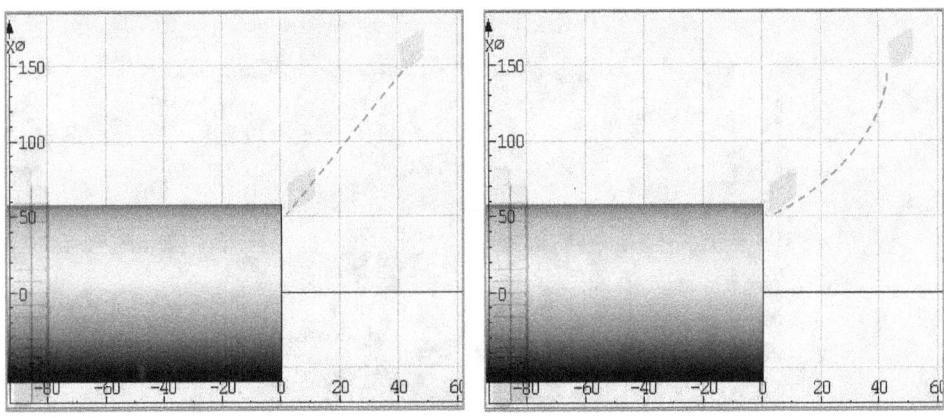

Fig. 85. Trajectory of rapid approach with the functions RTLION and RTLIOF

CNC – Basic Programming Course

11.2 G1: linear interpolation

Function G1 sets a working movement to be executed in linear interpolation. The programmed point is reached describing a straight line starting from the point where the tool is located.

The feed rate used is the active modal feed rate or it is specified in the same block. If in the destination block one of the two coordinates does not change, it is not necessary to enter it again; in this case the movement will be only on the programmed axis.

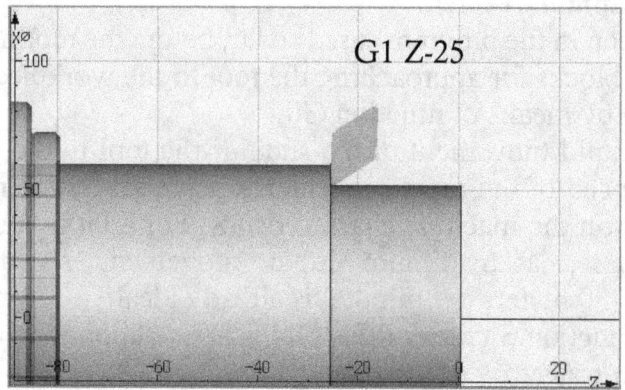

Fig. 86. Movement of the tool along the Z-axis

If, in the destination block, two values are programmed which differ from the starting point, the movement of the tool will occur along an inclined line, achieved by means of interpolation of the two axes (s. par. 4.8).

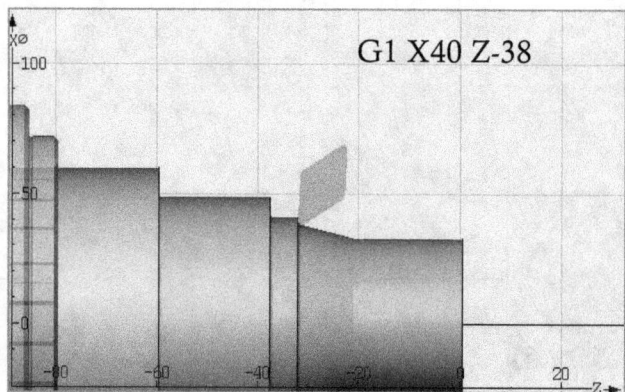

Fig. 87. Linear interpolation with tool moving along the axes X and Z

11.3 G33, G34, G35: threading in multiple passes

Function G33 sets a linear interpolation movement like the one set by G1, though synchronizing the start of the block with the angular zero position of the spindle.

This allows the execution of a threading with constant lead in multiple passes, during which the tool is always located in the same path, as shown in the next figure.

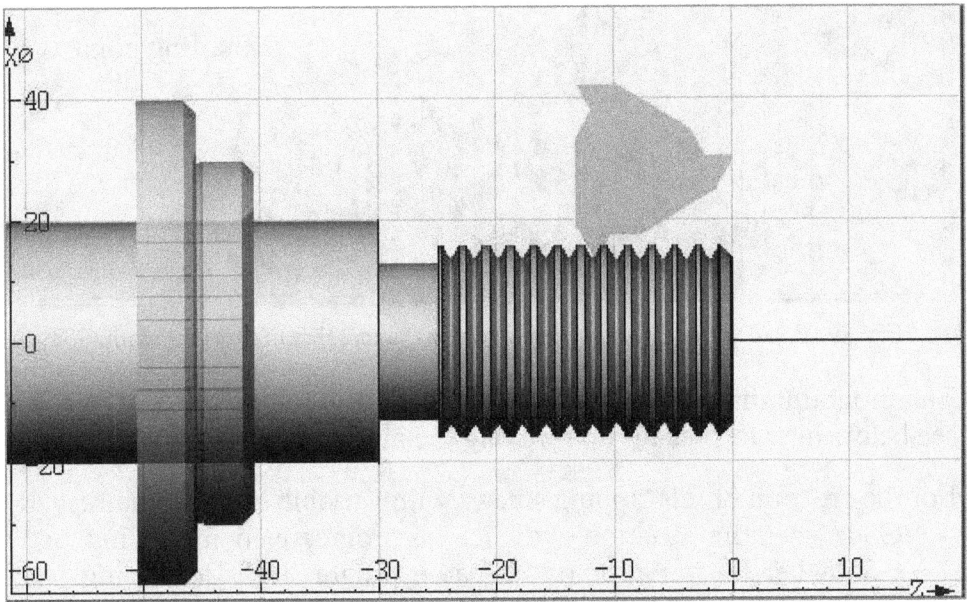

Fig. 88. Execution of a threading in multiple passes with G33

The lead of the thread is expressed in the same block as G33, using the address 'K' if the movement is on the Z-axis (G33 Z-20 K2), or 'I' (more unlikely) if the movement is on the X-axis.

The function G33 must always be programmed with a constant number of revolutions (G97/G95).

For conical threads, program the arrival point in X and Z and enter the value of the lead as a projection of the real lead on the predominant axis (Z for inclination angles under 45° and X for angles over 45°).

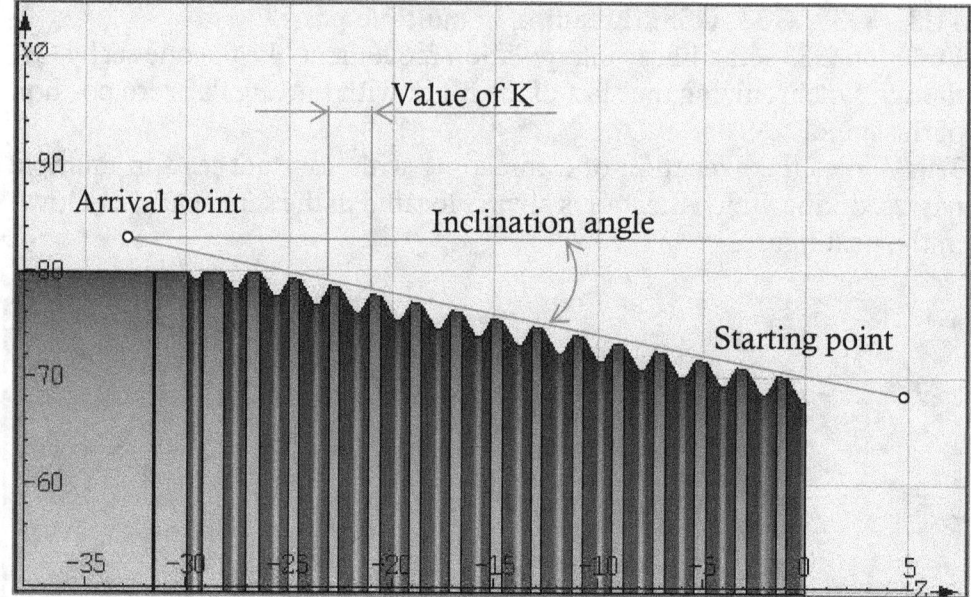

Fig. 89. Lead value to be programmed in a conical thread executed with G33

The programming of a thread with G33 is quite a long process. We will see below how to speed it up by using the automatic cycle CYCLE99.

For the creation of self-tapping screws with variable lead, program:
- G34: when the variation of the lead is progressively increasing, or
- G35: when the variation of the lead is progressively decreasing,

followed by the coordinates of the arrival point, the lead of the first spiral and the incremental value 'F' of the lead variation; for example

G34 Z-20 K2 F0.1

11.4   G4: dwell function

The function G4, when followed by the address 'F', sets a dwell time in seconds, or, if followed by the address 'S', in number of spindle revolutions. It is useful to guarantee the cylindricity of the bottom of grooves, in order to break or remove the chip during a drilling operation or in order to wait for a generic event (arrival of cooling liquid).

G4 F1   ; dwell time of 1 second
G4 S2   ; dwell time of 2 spindle revolutions

11.5 Practical exercise

11.5.1 Example of the roughing of a profile

Open the program PRG_11_01 in the folder CHAP_11_13, start the graphic simulation and enable the single block mode.
This program exclusively executes the roughing of the workpiece shown in the drawing below. Analyze the value of the programmed coordinates in every block and answer the questions in the paragraph below.

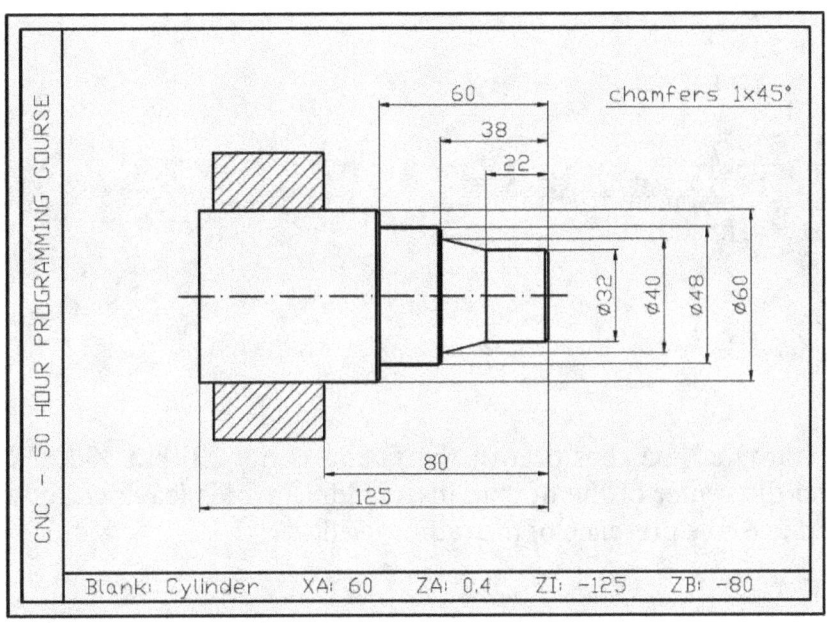

Fig. 90. Example of the programming of an external turning

```
; blank part dimensions:
; XA = 60 bar diameter
; ZA = 0.4 machining allowance on front face
; ZI = -125 length of finished part
; ZB = -80 extension from jaws
WORKPIECE(,,,"CYLINDER",0,0.4,-125,-80,60)

G18 G54 G90 ;G54 PART ZERO POINT SETTING
; G90 ABSOLUTE COORDINATE SYSTEM
G0 X400 Z500 ;SAFETY POSITION
M8 ; COOLANT ACTIVATION
SETMS(1) ; SETTING OF MASTER SPINDLE

T1 D1 ; ROUGHING TOOL SELECTION
```

```
G95 S740 M4 ; SETTING OF NUMBER OF REVOLUTIONS AND FEED RATE
IN MM/REV
G0 X52 Z2 ; RAPID APPROACH TO WORKPIECE
G1 Z-59.8 F0.2 ;FIRST PASS WITH FEED RATE OF 0.2 MM/REV
G0 X54 Z2 ; RETURN OUTSIDE OF THE WORKPIECE FACE
G0 X46 ; RAPID POSITIONING AT DIAMETER 46
G1 Z-37.8 ;SECOND PASS
G1 X49
G1 Z-59.8
G0 X51 Z2
G0 X41 ;THIRD PASS
G1 Z-37.8
G0 X43 Z2
G0 X33 ;FOURTH PASS
G1 Z-21.8
G1 X41 Z-37.8
G0 Z0 ; POSITIONING FOR FACING
G0 X35 ; APPROACH TO DIAMETER
G1 X-1.6 ;EXECUTION OF FACING
G0 X60 Z2

G0 X200
G0 Z200
M30
```

Note how the final position of the facing is not X0 but X-1.6. Going beyond the center of the double insert radius avoids leaving the witness mark due to the presence of the radius itself.

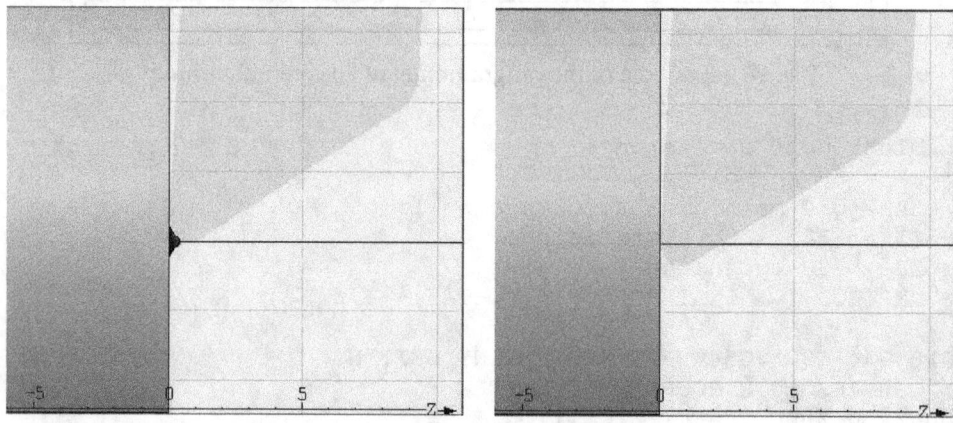

Fig. 91. Elimination of the facing witness mark

11.5.2 Review of program comprehension
Answer the following questions about the program you've just executed.

1) What is the cutting speed of tool T1 during the first pass?
   a) 120 b) 90 c) 100

2) What is the radial depth in X of the first pass?
   a) 5 b) 8 c) 4

3) At what value is diameter 48 roughed?
   a) 52 b) 49 c) 59.8

4) How much allowance is left on the shoulder at Z-38?
   a) 0.2 b) 0.1 c) 59.8

5) What are the coordinates of the position where the tool is located before executing the facing?
   a) X35, Z0 b) X41, Z-0.4 c) X62, Z0

The correct solutions can be found in the program ANS_11_01 in the folder CHAP_11_13.

11.5.3 Example of the programming of a threading

Open the program PRG_11_02 in the folder CHAP_11_13, start the graphic simulation and enable the single block mode.

This program executes:
- with tool T1 D1, the facing, the chamfer and the external turning of the workpiece,
- with tool T3 D1 (with 3 millimeter insert) the groove at the end of the thread,
- with tool T4 D1, the thread in multiple passes.

The number of passes in order to execute a thread depends on the dimensions of its lead. The depth of every pass is recommended by the tools' manufacturer according to the material to be worked on.
In this case, 8 passes have been executed, the depth of which is shown in the program.

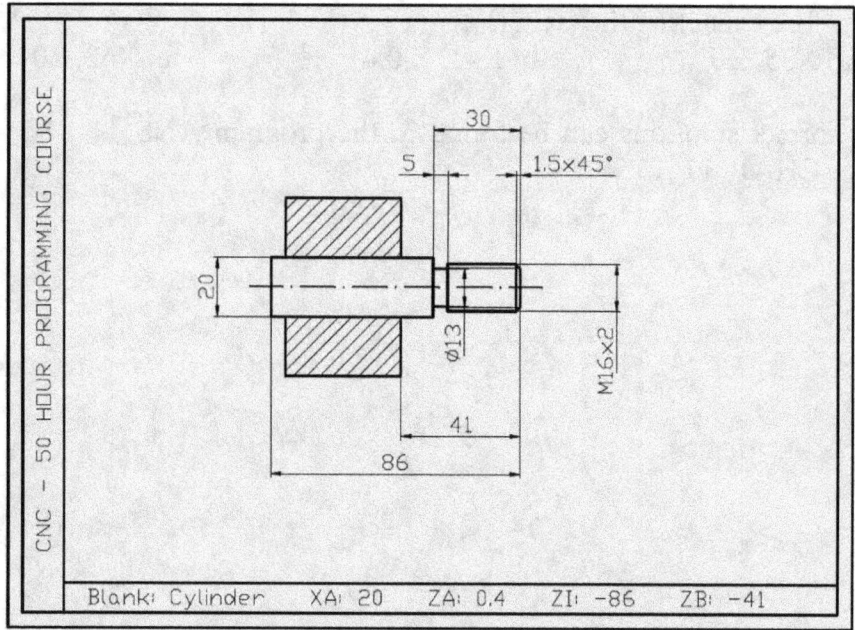

Fig. 92. Example of the programming of a threaded workpiece

```
; blank part dimensions:
; XA = 20 bar diameter
; ZA = 0.4 machining allowance on front face
; ZI = -86 length of finished part
; ZB = -41 extension from jaws
```

```
WORKPIECE(,,,"CYLINDER",0,0.4,-86,-41,20)

G18 G54 G90 ;G54 PART ZERO POINT SETTING
; G90 ABSOLUTE COORDINATE SYSTEM
G0 X400 Z500 ;SAFETY POSITION
M8 ; COOLANT ACTIVATION
SETMS(1) ; SETTING OF MASTER SPINDLE

LIMS=3000 ; MAXIMUM LIMIT OF REVOLUTIONS
T1 D1 ; EXTERNAL TURNING
G96 S100 M4 ; SETTING OF CONSTANT CUTTING SPEED AND FEED RATE
IN MM/REV
G0 X22 Z0 ; RAPID APPROACH TO WORKPIECE
G1 X-1.6 F0.18 ; FACING
G0 X12.8 Z0.5 ; DIAMETER AT THE BEGINNING OF THE CHAMFER
G1 Z0 ; APPROACH TO FACE OF WORKPIECE
G1 X15.8 Z-1.5 ; EXECUTION OF THE CHAMFER 1.5 X 45
G1 Z-30 ; TURNING
G1 X22 ; STRAIGHT SHOULDER
G0 X200 ; DISENGAGEMENT IN X
G0 Z200 ; DISENGAGEMENT IN Z

T3 D1 ; GROOVING TOOL WIDTH 3MM
G95 S800 M4 ; SETTING OF NUMBER OF REVOLUTIONS AND FEED RATE
IN MM/REV
G0 Z-30 ; RAPID POSITIONING IN Z
G0 X22 ; APPROACH TO DIAMETER OF BAR
G1 X13 F0.1 ; EXECUTION OF THE GROOVE
G4 S2 ; DWELL TIME OF TWO REVOLUTIONS AT BOTTOM OF GROOVE
G0 X22
G0 Z=IC(2) ; INCREMENTAL MOVEMENT OF 2MM IN Z POSITIVE
G1 X13
G4 S2
G0 X22
G0 X200 ; DISENGAGEMENT IN X
G0 Z200 ; DISENGAGEMENT IN Z

T4 D1 ; TOOL FOR EXTERNAL THREADS
G95 S600 M3 ; INVERSION OF SPINDLE ROTATION DIRECTION
G0 Z4 ; RAPID POSITIONING IN Z

; POSITIONING AT DIAMETER OF FIRST PASS BEGINNING FROM A NOMINAL
DIAMETER OF 16 MM
G0 X15.4 ; DEPTH OF RADIAL PASS OF 0.3MM
G33 Z-29.5 K2
G0 X18 ; EXIT FROM THREAD
G0 Z4 ; REPOSITIONING IN Z
```

```
G0 X14.9 ; DEPTH OF RADIAL PASS OF 0.25MM
G33 Z-29.5 K2 ; SECOND PASS
G0 X18 ; EXIT FROM THREAD
G0 Z4 ; REPOSITIONING IN Z

G0 X14.5 ; DEPTH OF RADIAL PASS OF 0.2MM
G33 Z-29.5 K2 ; THIRD PASS
G0 X18 ; EXIT FROM THREAD
G0 Z4 ; REPOSITIONING IN Z

G0 X14.1 ; DEPTH OF RADIAL PASS OF 0.2MM
G33 Z-29.5 K2 ; FOURTH PASS
G0 X18 ; EXIT FROM THREAD
G0 Z4 ; REPOSITIONING IN Z

G0 X13.8 ; DEPTH OF RADIAL PASS OF 0.15MM
G33 Z-29.5 K2 ; FIFTH PASS
G0 X18 ; EXIT FROM THREAD
G0 Z4 ; REPOSITIONING IN Z

G0 X13.56 ; DEPTH OF RADIAL PASS OF 0.12MM
G33 Z-29.5 K2 ; SIXTH PASS
G0 X18 ; EXIT FROM THREAD
G0 Z4 ; REPOSITIONING IN Z

G0 X13.36 ; DEPTH OF RADIAL PASS OF 0.10MM
G33 Z-29.5 K2 ; SEVENTH PASS
G0 X18 ; EXIT FROM THREAD
G0 Z4 ; REPOSITIONING IN Z

G0 X13.26 ; DEPTH OF RADIAL PASS OF 0.05MM
G33 Z-29.5 K2 ; EIGHTH PASS
G0 X18 ; EXIT FROM THREAD

G0 X200 ; DISENGAGEMENT
G0 Z200
M30
```

11.5.4 Finishing of a profile
This exercise allows for the consolidation of the learning of much of the information delivered up to now.
- Open the program PRG_11_01 in the folder CHAP_11_13
- copy it into the folder 01_EXERCISES according to the procedures laid down in paragraph 4.10.3
- change the name into EX_11_01
- **at the end of the program, after the roughing, enter the finishing of the profile according to the programming sequence specified in paragraph 5.2**
- the part to be created is described in the drawing in fig. 90
- make sure the starting values of the chamfers in X (paragraph 4.10.1) are correctly programmed

Compare your program to the one in the folder FINISHED_EXERCISES named EX_11_01.

12. Direct Programming of Rounds, Chamfers and Angles (2h)
(Theory: 1h, Practice: 1h)

12.1 Introduction
By now, we've seen that all segments constituting a profile are defined by programming the coordinates of their arrival point.
There is also a simplified programming method, which delegates to the NC the calculation of the tool path by directly programming rounds, chamfers and inclination angles of the lines compared to the main rotating axis.

12.2 RND= / RNDM=: execution of a round
The function RND, followed by the value of the radius, allows for the entry of a tangential round between linear and circular parts of the profile at the end of a block.

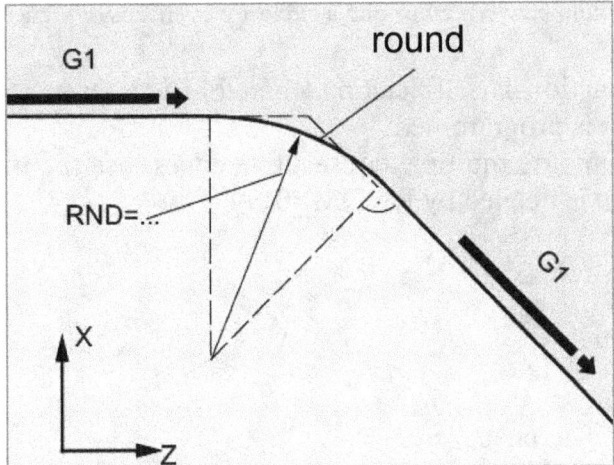

Fig. 93. Round between two line by means of the RND function

The starting and arrival point of the round depend on the dimensions of the programmed radius and the direction of the two blocks to be connected.
RND is not a function designed for the programming of a circle arc; it is instead designed to simplify the programming of the breaking of a sharp edge by a round radius.

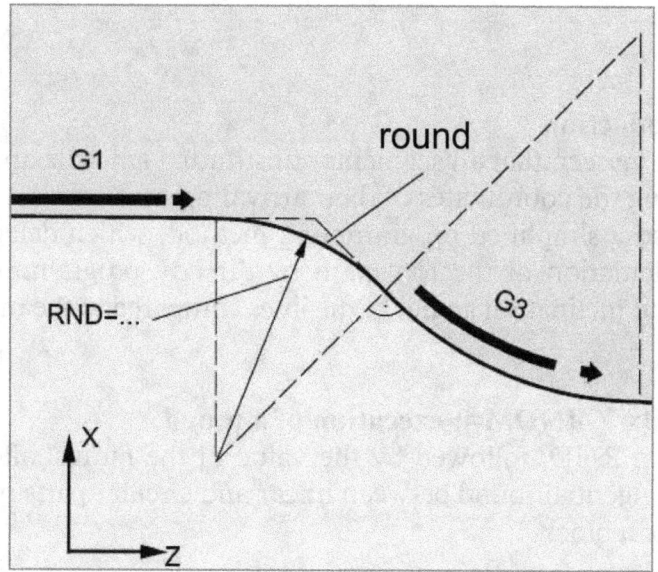

Fig. 94. Round between a line and a circle arc by means of the RND function

The RND function is self-deleting, therefore it is executed only in the block where it is programmed.
In order to combine multiple consecutive edges, use the modal function RNDM which is deleted by RNDM=0.

Programming syntax:

```
G1 Z-20 RND=0.4
G1 X18 Z-14
```

or:

```
G1 Z-20 RNDM=0.4
G1 X18 Z-14
```

12.3 CHR= / CHF=: execution of a chamfer

As for the function RND, the function CHR, followed by the dimensions of the chamfer, allows for the entry of the edge breaking located at the intersection between linear or circular parts of the profile at the end of a block.

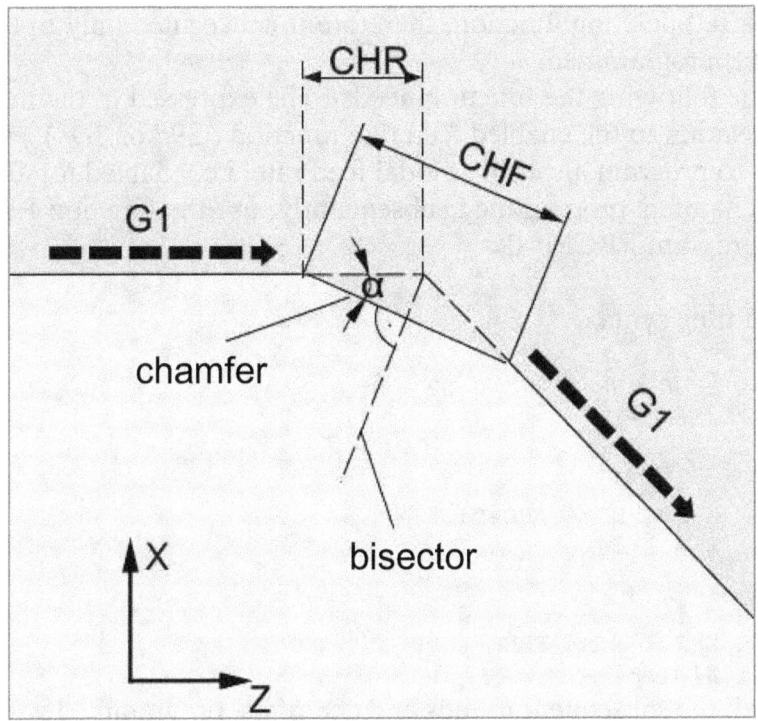

Fig. 95. Chamfer executed between two lines by means of the function CHR or CHF

The dimension of the chamfer is defined with CHR when the value is expressed according to the direction of the segments, or with CHF when the value refers to the real length of the chamfer.

The starting and the arrival point of the chamfer depend on the type of programmed function (CHR or CHF), on the dimensions of the chamfer and on the direction of the two profile blocks.

CHR and CHF are not functions designed for the programming of an inclination angle of the chamfer; they are instead designed to simplify the programming of the breaking of a sharp edge.

12.4 FRC= / FRCM: specific feed rate on chamfers and rounds

In order to optimize the surface quality of chamfers and rounds, it is possible to program, by means of the function FRC (Feed rate on Round and Chamfer), a specific feed rate with which to create them.

FRC, followed by the feed rate value, needs to be entered in the same block in which you program radii or rounds.

FRC is a self-deleting function, therefore it is executed only in the block where it is programmed.

The value following the function needs to be expressed in the measuring unit in relation to the enabled feed rate function (G95 or G94).

In order to program a specific modal feed rate, i.e. enabled for all rounds and/or chamfers programmed subsequently, use the function FRCM; to disable program FRCM=0.

Programming syntax:

```
G1 Z-20 CHR=1 FRC=0.02
G1 X18 Z-14
```

or:

```
G1 Z-20 RND=4 FRC=0.02
G1 X18 Z-14
```

or:

```
G1 Z-20 RND=4 FRCM=0.02
G1 X18 Z-14
```

(all the subsequent rounds and chamfers, programmed with RND, CHR and CHF, are executed with the specific feed rate of 0.02 millimeters per revolution)

12.5 ANG=: direction of a line defined by an angle

The function ANG allows for the definition of profile lines by directly programming the inclination angle value of the line with respect to the positive direction of the Z-axis.

The value of the angle to be programmed is obtained by using the scheme below, already presented in paragraph 4.9.

Position the tool tip at the center of the Cartesian axes; the direction of the path you want to execute indicates the inclination angle to be programmed.

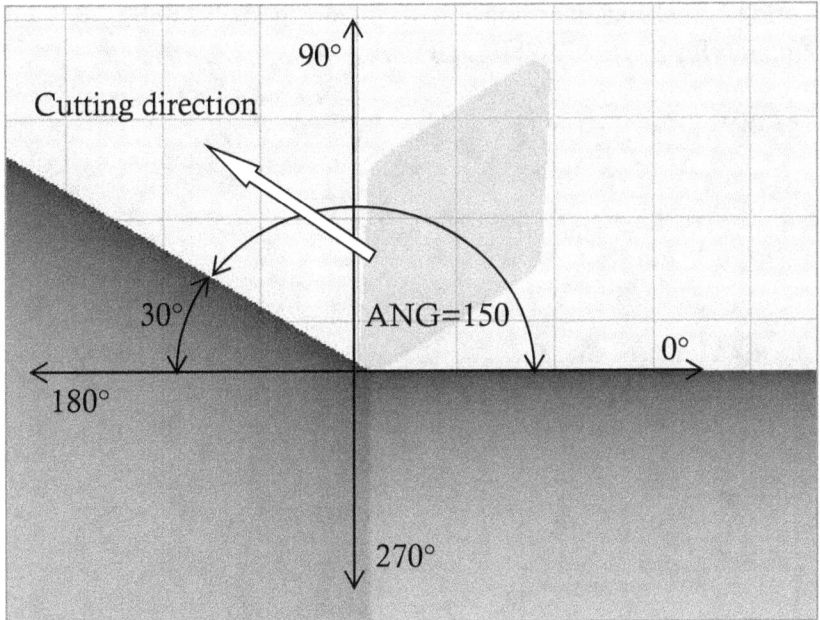

Fig. 96. Scheme for the definition of the angle by means of the function ANG

The block describing the line must contain one single coordinate of the arrival point (either X or Z) and the inclination angle with which to reach it.

Programming syntax: G1 Z-40 ANG=150

It is furthermore possible to program in a block only the cutting direction and in the next block the coordinates of the arrival point in X and Z together with the angle value.

Programming syntax: G1 ANG=180
                      G1 Z-38 X40 ANG=166

The arrival point of the first block is calculated by the NC on the basis of the position of the second point and the direction of the two lines.

At the end of the block it is possible to program radii or chamfers with the functions RND, RNDM, CHR, CHF.

12.6 Practical exercise

12.6.1 Point to point and direct programming comparison

In paragraph 11.5.4 we executed the finishing program of the profile by always entering the coordinates of the arrival point.
Now, the same drawing will be used, the cone though is defined by means of only the arrival point in Z together with its inclination angle.

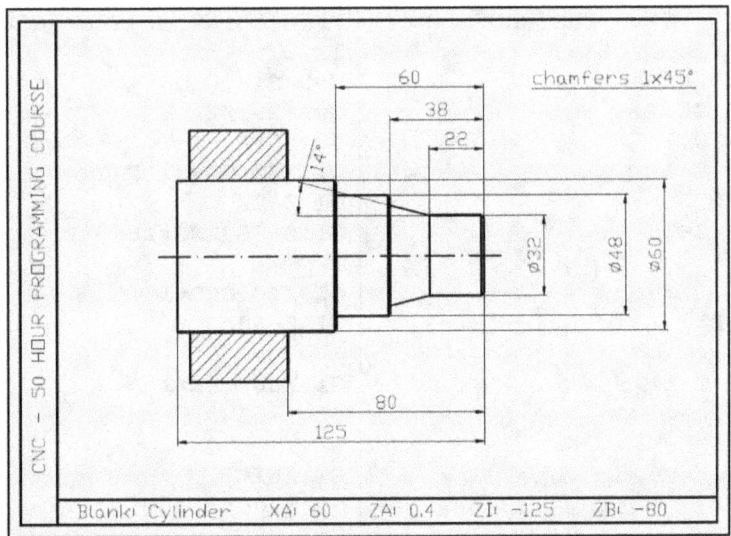

Fig. 97. Programming of a profile by means of the functions CHR, FRCM and ANG

Open the program PRG_12_01 in the folder CHAP_11_13.
In this program no changes have been made to the roughing path, but the finishing is programmed using the direct programming functions for chamfers and angles.
Start the graphic simulation and enable the single block mode: analyze the programmed functions and the respective tool movement.

Compare the new program on the next page with the previous one.
You will see that for the execution of the frontal chamfer it is necessary to change the starting point and program a vertical line which intersects with the next block at an angle of 90°.

The value of the inclination angle to be programmed for the execution of the cone refers to the positive direction of the Z-axis as shown in figure 96 and in the programming scheme in paragraph 4.9.

Previous program created by programming the arrival point coordinates.	New program created by the direct programming of chamfers and angles.
;FINISHING OF THE PROFILE T2 D1 G95 S1800 M4 **G0 X30 Z2** G1 Z0 F0.1 G1 X32 Z-1 G1 Z-22 G1 X40 Z-38 G1 X46 G1 X48 Z-39 G1 Z-60 G1 X58 G1 X60 Z-61 G1 Z-62 G1 X61 G0 X200 G0 Z200 M30	;FINISHING OF THE PROFILE T2 D1 G95 S1800 M4 **G0 X26 Z2** G1 Z0 F0.1 **G1 X32 CHR=1 FRCM=0.04** G1 Z-22 **G1 Z-38 ANG=166** **G1 X48 CHR=1** G1 Z-60 **G1 X60 CHR=1** G1 Z-62 G1 X61 G0 X200 G0 Z200 M30

Fig. 98. Comparison between the two programs which create the same profile: in the left column by means of the point to point programming, in the right column by using the direct programming functions CHR, FRCM and ANG

In order to use the direct programming functions for chamfers and rounds it is recommended that the block where they are programmed be followed by a working movement and not by a rapid movement.

The space defined between the starting block and the arrival block must be sufficient to contain the dimensions of the programmed chamfer or round.

12.6.2 Definition of the blank part data

The dimensions of the blank part are used by the graphic simulation in order to display the part to be worked on.
The blank part data need to be entered at the beginning of the program.

Before that, it is recommended to write down comments which report their dimensions, as has been done in all the programs used until now.

Following the procedure specified in paragraph 8.10.2, create a new main program (.MPF) in the folder 01_EXERCISES and name it EX_12_01.
At the beginning, the program is empty. Enter the comments which describe the dimensions of the blank part with reference to the drawing in figure 100.

```
; blank part dimensions:
; XA = 50 bar diameter
; ZA = 0.3 machining allowance on front face
; ZI = -100 length of finished part
; ZB = -70 extension from jaws
```

The table of blank data shown by the NC identifies the basic parameters for its definition with the letters XA, ZA, ZI and ZB. Their meaning is specified in paragraph 3.4.

Place the cursor on the line after the comments in order to enter the blank data.

Press the horizontal softkey VARIOUS.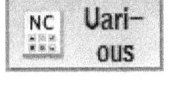

Then press the vertical softkey BLANK.

Enter the values and confirm with ACCEPT.

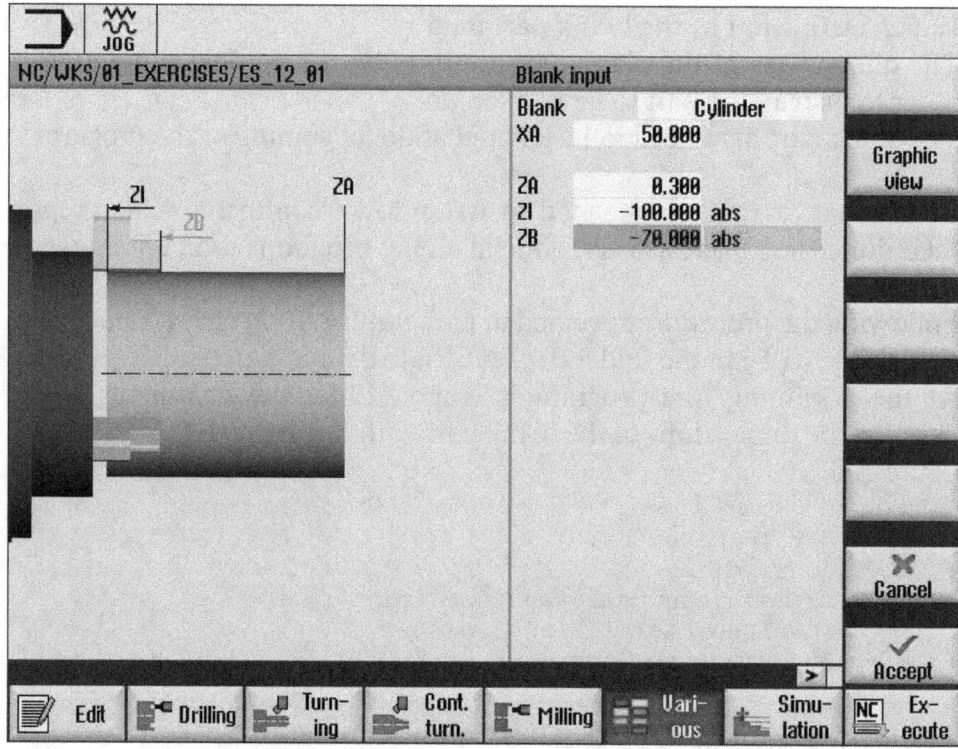

Fig. 99. Page for entering the blank part data

In order to return and modify the values after their acceptance, press the arrow shown at the end of the block with your mouse.

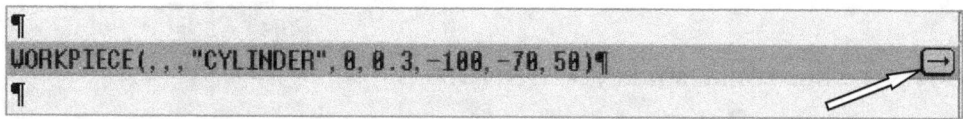

Attention: with the SELECT button it is possible to set the value of ZI and ZB with absolute or incremental meaning. The selection of 'absolute' shows that the value expressed refers to the part zero point; the selection of 'incremental' shows that the value expressed refers to the front face of the workpiece inclusive of machining allowance set at parameter ZA.

12.6.3 Programming of a workpiece

Enter the missing data into the program below.
The arrow (→) before the block number tells you where to enter the value.
After entering the values, write the complete text in the newly created program EX_12_01.

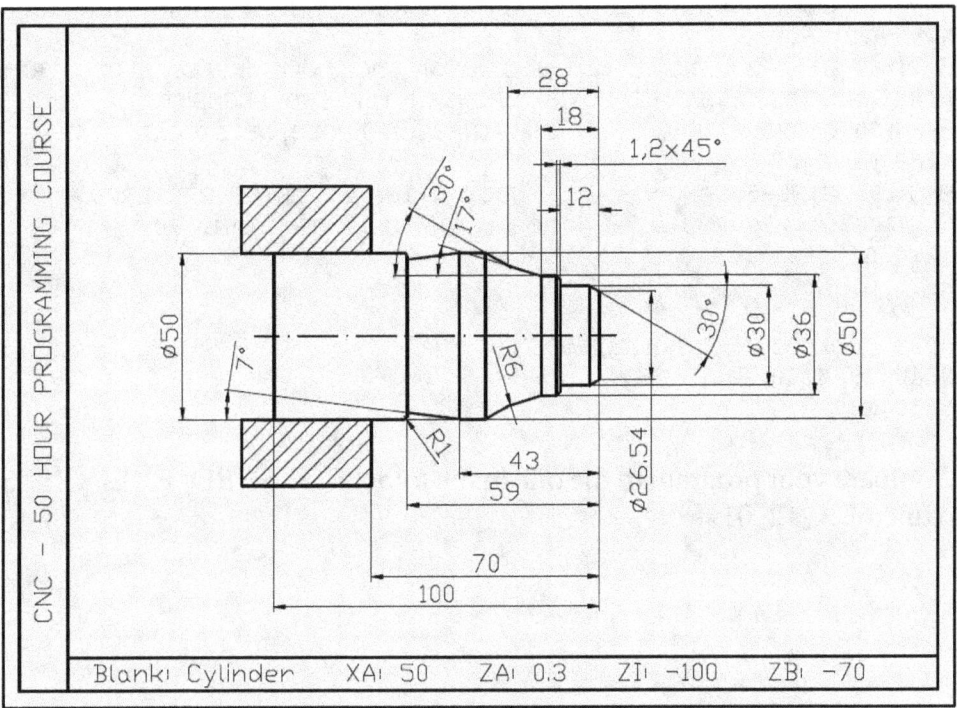

Fig. 100. Enter the missing data for the execution of this profile

```
; blank part dimensions:
; XA = 50 bar diameter
; ZA = 0.3 machining allowance on front face
; ZI = -100 length of finished part
; ZB = -70 extension from jaws
N10 WORKPIECE(,,,"CYLINDER",0,0.3,-100,-70,50)

N20 G18 G54 G90
N30 G0 X400 Z500
N40 M8
N50 SETMS(1)

N60 LIMS=2200
```

```
N70 T1 D1
N80 G96 S100 M4
→ N90 G0 X52 Z............ ; POSITIONING FOR FACING
N100 G1 X-1.6 F0.18
→ N110 G0 X............ Z2 ; STARTING DIAMETER OF CHAMFER AT 30°
N115 G1 Z0
→ N120 G1 X30 ANG=............ ; ANGLE FOR THE EXECUTION OF THE CHAMFER
N130 G1 Z-12
→ N140 G1 X36 CHR=............ ; DIMENSION OF CHAMFER AT 45°
N150 G1 Z-18
→ N160 G1 Z-28 ANG=............ RND=............ ; INCLINATION ANGLE OF FIRST
LINE AND VALUE OF THE ROUND BETWEEN THE TWO SEGMENTS
→ N170 G1 X50 ANG=............ ; INCLINATION ANGLE OF SECOND LINE
N180 G1 Z-43
→ N190 G1 Z-59 ANG=............ ; INCLINATION ANGLE OF THE LINE
→ N200 G1 X50 RND=............ ; ROUND RADIUS WITH DIAM. 50
N210 G1 Z-61
N220 G1 X51

N230 G0 X200 ; DISENGAGEMENT
N240 G0 Z200
N250 M30
```

Compare your program to the one in the folder FINISHED_EXERCISES named EX_12_01.

13. Circular Interpolation (1h)
(Theory: 0.5h, Practice: 0.5h)

13.1 G2: circular interpolation in clockwise direction

The function G2 allows for the programming of circle arcs traveled by the tool in clockwise direction. The clockwise direction is defined according to the programming scheme shown in paragraph 4.9.

The circle arc is programmed by entering the function G2, followed by the coordinates of the arrival point and the radius dimensions (CR=).

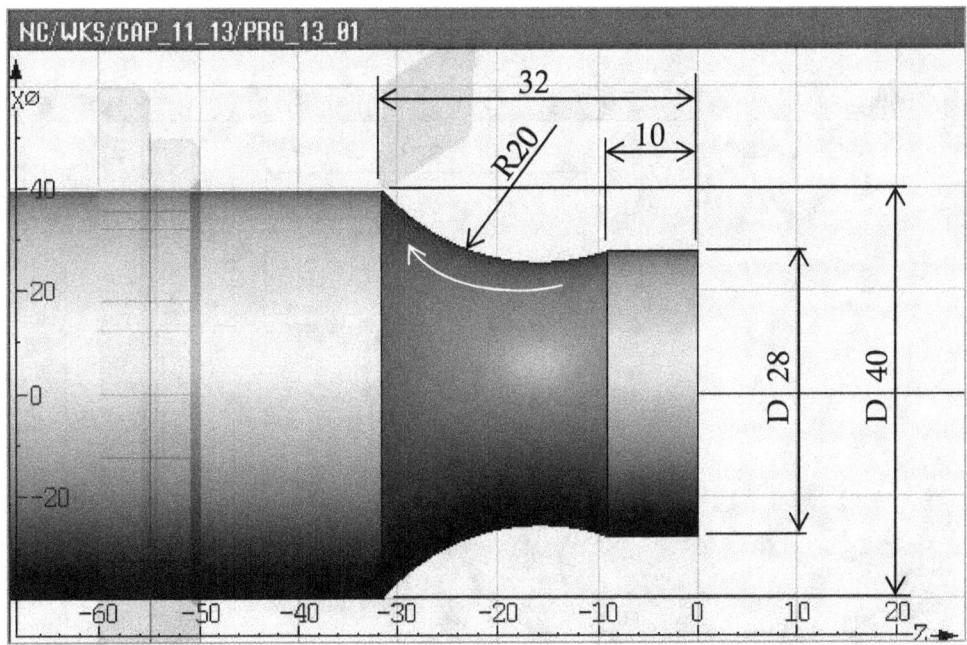

Fig. 101. G2 : circular interpolation in clockwise direction

Below you will find a programming example for the creation of the profile shown in figure 101:

```
N10  WORKPIECE(,,,"CYLINDER",0,0,-80,-50,40)
N20  G18 G54 G90
N30  G0 X400 Z500
N40  M8
N50  SETMS(1)
N60  T1 D1 ; TURNING TOOL
N70  G95 S1400 M4
N80  G0 X28 Z2
N90  G1 Z-10 F0.18
N100 G2 X40 Z-32 CR=20
N110 G1 X41
N120 G0 X200
N130 G0 Z200
N140 M30
```

13.2 G3: circular interpolation in counterclockwise direction

The function G3 allows for the programming of circle arcs traveled by the tool in counterclockwise direction.

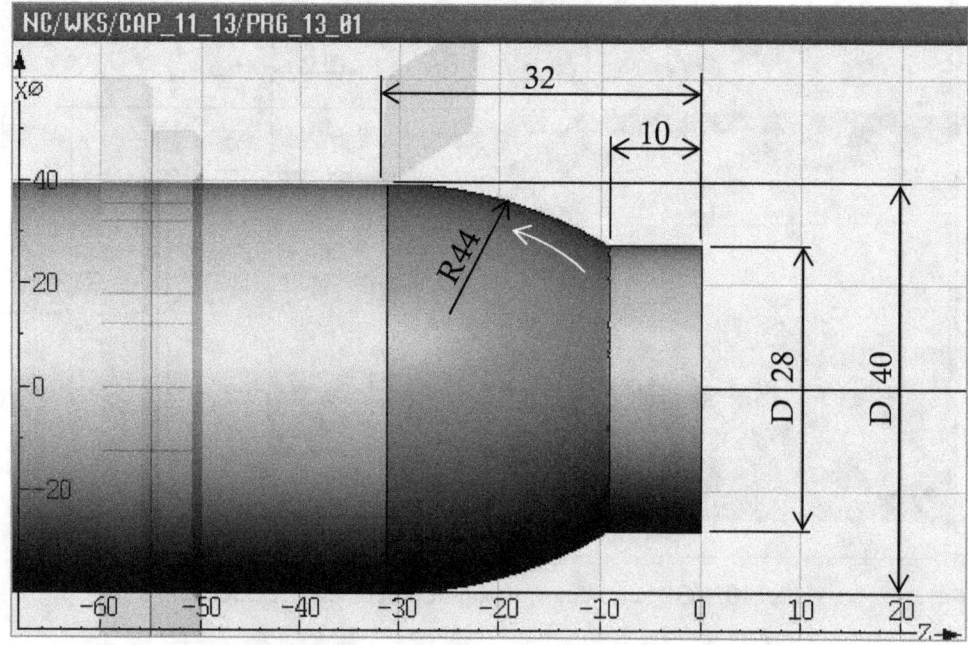

Fig. 102. G3 : circular interpolation in counterclockwise direction

Below you will find a programming example for the creation of the profile shown in figure 102:

```
N10  WORKPIECE(,,,"CYLINDER",0,0,-80,-50,40)
N20  G18 G54 G90
N30  G0 X400 Z500
N40  M8
N50  SETMS(1)
N60  T1 D1 ; TURNING TOOL
N70  G95 S1400 M4
N80  G0 X28 Z2
N90  G1 Z-10 F0.18
N100 G3 X40 Z-32 CR=44
N110 G1 X41
N120 G0 X200
N130 G0 Z200
N140 M30
```

13.3 I, K, I=AC(...), K=AC(...): progr. of the radius center

In the previous paragraphs, the circle arc has been defined by programming its arrival point and the radius value.

Another option is to program the coordinates of the radius center on X and Z (or Y when available) instead of the radius.

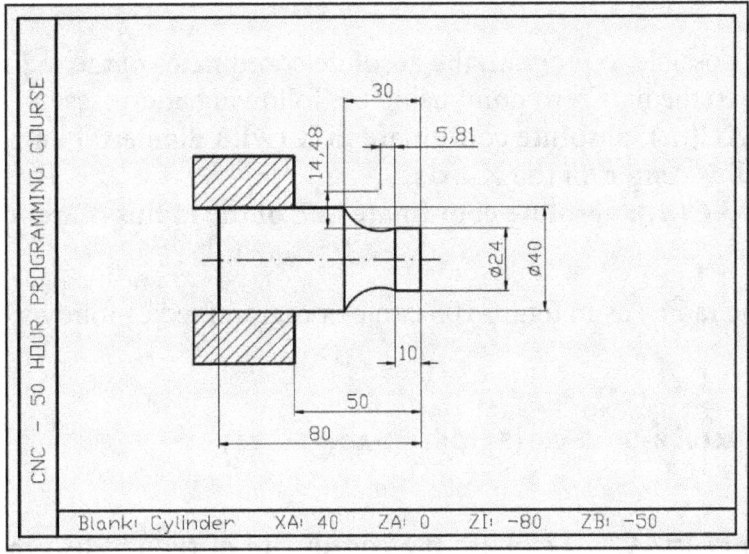

Fig. 103. Programming of an arc by means of the radius center coordinates

These coordinates may be expressed as incremental values referring to the starting point of the arc using the addresses I and K.

I: expresses the coordinate of the radius center with respect to the starting point of the arc on the X-axis (with radial value).
K: expresses the coordinate of the radius center with respect to the starting point of the arc on the Z-axis.

The following program executes the profile shown in figure 103:

```
N10  WORKPIECE(,,,"CYLINDER",0,0,-80,-50,40)
N20  G18 G54 G90
N30  G0 X400 Z500
N40  M8
N50  SETMS(1)
N60  T1 D1 ; TURNING TOOL
N70  G95 S1400 M4
N80  G0 X24 Z2
N90  G1 Z-10 F0.18
N100 G2 X40 Z-30 I14.48 K-5.81
N110 G1 X41
N120 G0 X200
N130 G0 Z200
N140 M30
```

It is also possible to program the absolute coordinates of the radius center referring to the part zero point using the following addresses:

I=AC(...), absolute coordinate in X (with diametral value) of the radius center on the X-axis.
K=AC(...): absolute coordinate in Z of the radius center on the Z-axis.

The same radius as in figure 103 can be programmed as follows:

```
N80  G0 X24 Z2
N90  G1 Z-10 F0.18
N100 G2 X40 Z-30 I=AC(52.96) K=AC(-15.81)
N110 G1 X41
```

J and J=AC(...) express the coordinate of the radius center with respect to the starting point of the arc on the Y-axis (plane G19).

13.4 Definition of the working plane

During turning operations, the tool always moves on the X-Z plane. When there is a Y-axis in the machine (see par. 4.5), there are two more working planes (X-Y and Z-Y), which are only used for milling operations.

The functions for circular interpolation require the programming of the working plane of the tool before their execution.

> G18 defines the working plane Z-X
> G19 defines the working plane Y-Z
> G17 defines the working plane X-Y

In a lathe, the X-Z plane is normally already enabled at machine start. As long as turning operations are carried out, it is therefore unnecessary to reprogram it; amongst the first blocks the function G18 was entered again, only as a reminder of the presence of this instruction. The functions G17 and G19 are programmed before the milling operations performed on these planes.

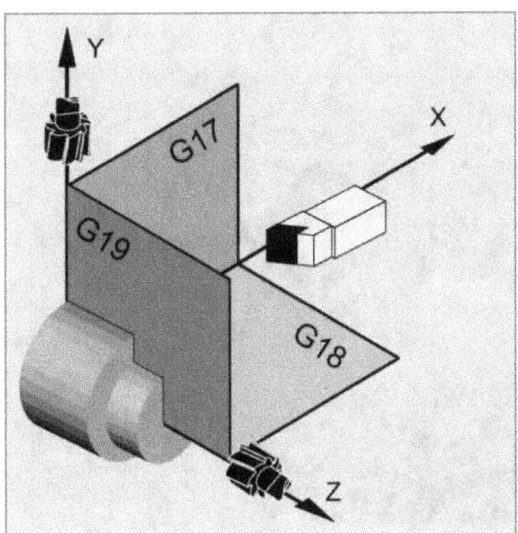

Fig. 104. Functions for the definition of the working plane

The definition of the working plane is essential, not only for the circular interpolations, but also for the tool radius compensation functions and for the direct programming of the angles.

13.5 Practical exercise

13.5.1 Programming of different radii
Open the program PRG_13_01 in the folder CHAP_11_13.
Copy it into the folder 01_EXERCISES and rename it as EX_13_01.
This program contains the execution of an arc in block N100.
Replace block N100 with the ones specified subsequently, start the graphic simulation, enable the single block mode and observe the programmed tool path.

In order to break the sharp edge at the end of the arc by means of a round, program RND= in the block of G2; remember to check and if necessary modify (as in this case) the direction of the next block to which it is joined.

```
N90 G1 Z-10 F0.18
N100 G2 X40 Z-32 CR=20 RND=4
G1 Z-36
N110 G1 X41
```

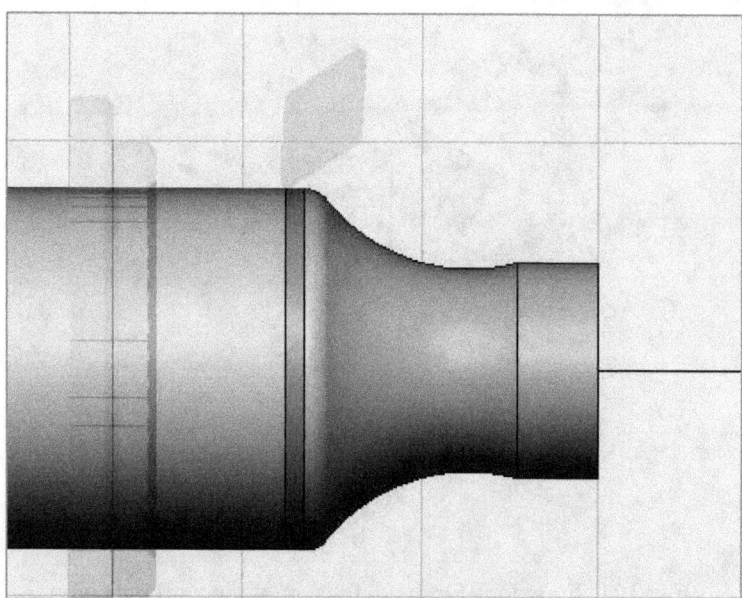

Fig. 105. Round between a radius and the following line by means of G2 and RND

In order to break the sharp edge at the end of the arc by means of a chamfer, program CHR= or CHF= in the block of G2 and, as before, remember to check the direction of the next block with which the arc is intersected.

```
N90 G1 Z-10 F0.18
N100 G2 X40 Z-32 CR=20 CHR=5
G1 Z-34
N110 G1 X41
```

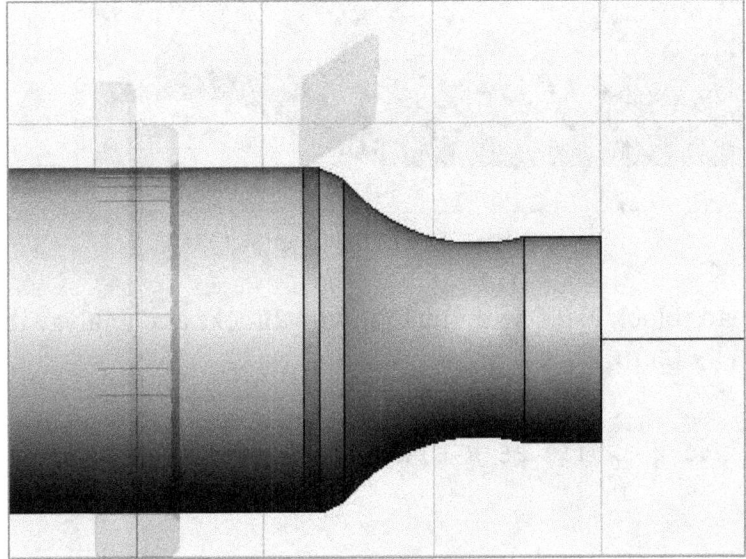

Fig. 106. Chamfer between a radius and the following line by means of G2 and CHR

Now use the function G3 and note how the shape of the programmed radius changes.

```
N90 G1 Z-10 F0.18
N100 G3 X40 Z-32 CR=40
N110 G1 X41
```

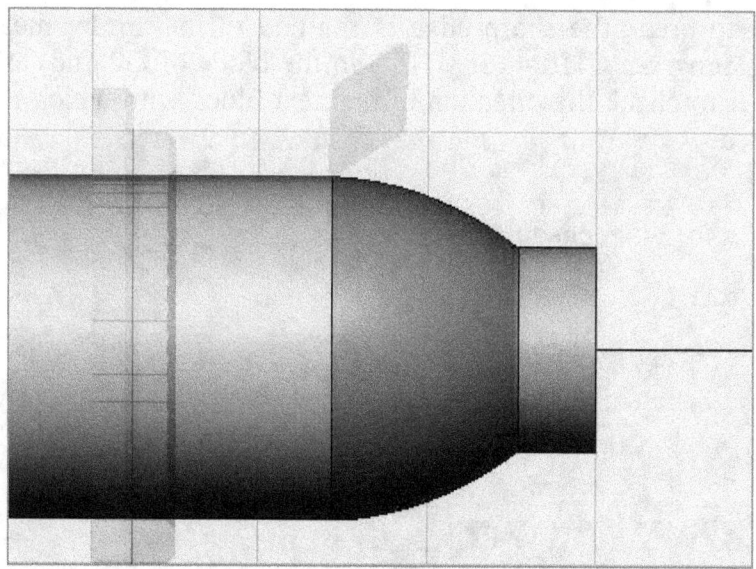

Fig. 107. Use of the function G3

Now replace block N100 with the following blocks and analyze the profile described by them.

```
N90  G1 Z-10 F0.18
N100 G2 X40 Z-32 I18.35 K-6.81
N110 G1 X41

N90  G1 Z-10 F0.18
N100 G2 X40 Z-32 I=AC(64.7) K=AC(-16.81)
N110 G1 X41
```

Try programming different radii by changing the coordinates of the arrival point and the radius dimensions. If there are radii and points which are not correctly programmed, alarms will be displayed.

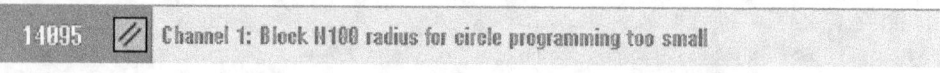

Fig. 108. Type of alarm displayed in the event of a programming error in a radius

14. First Test (2h)
(Practice: 2h)

14.1 Introduction to the test

The test consists in the execution of the program creating the part shown in figure 111. Take the following steps:
- Load the tool files contained in folder 01_EXERCISES named EMPTY_TOOL_LIST. This file deletes all the existing tools by overwriting them only with the roughing tool defined therein. In order to load this file, follow the procedure laid down in paragraph 3.3.
- Now create the necessary tools for the execution of this program following the procedure specified in paragraph 7.5.1. Below is a list of the necessary tools, their position in the turret, the offset data in X and Z and the data for the definition of their graphic aspect.

Loc.	Type	Tool name	ST	D	Length X	Length Z	Ø		Tip angle	
1		ROUGHING TOOL	1	1	88.000	40.000	0.800 ←		93.0 55	11.0
2		FINISHING TOOL	1	1	94.000	40.000	0.200 ←		93.0 55	11.0
3		OD GROOVING W.3MM	1	1	98.000	40.000	0.100		3.000	10.0
4		OD THREADING	1	1	88.000	46.000	0.200			
5		CENTER DRILL D.6	1	1	100.000	24.000	6.000		118.0	
6		AX. DRILL D.8.5	1	1	100.000	56.000	8.500		118.0	

Fig. 109. List of tools to be created and used in the test program

- Create an empty main program in the folder 01_EXERCISES and name it TEST_14_01.
- Structure the program as the ones we've seen so far:

- Enter the comments with the dimensions of the blank part at the beginning of the program. If you want to copy blocks from already existing programs see paragraph 14.4.
- Define the dimensions of the blank part according to the procedure laid down in paragraph 12.6.2.
- Enter the blocks that activate the initial settings and the safety position:
  ```
  G18 G54 G90
  G0 X400 Z500
  M8
  SETMS(1)
  ```
- Proceed to the programming of the tooling operations following the logical sequence described in paragraph 5.2.

14.2 Tooling operations and cutting parameters

Tooling sequence	Tool	Operation	Cutting speed (m/min)	Feed rate (mm/rev)
1 st	T1 D1	Roughing	100	0.18
2 nd	T2 D1	Finishing	120	0.12
3 rd	T3 D1	Groove	78	0.1
4 th	T4 D1	Threading	60	-
5 th	T5 D1	Center drilling	80	0.08
6 th	T6 D1	Hole D8.5	80	0.1

Fig. 110. Sequence of tooling operations and cutting parameters to use in the test

14.3 Drawing of the part to create

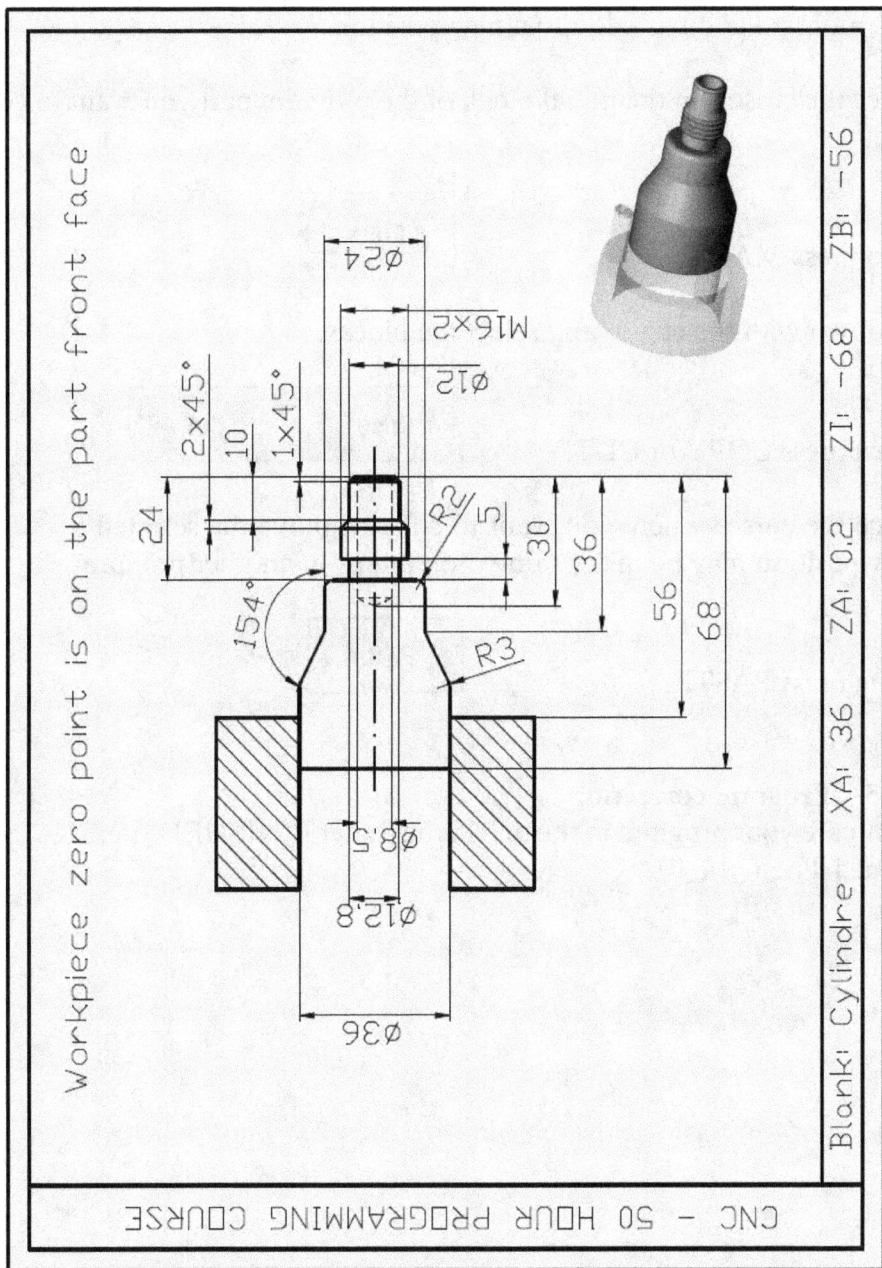

Fig. 111. Drawing of the part to create

14.4 Copying & pasting of program parts

In order to speed up programming times it is possible to copy or cut part of a program and paste it into a new position.

Place the cursor on the initial block of the program part you want to copy or cut,

then press MARK,

go down with the cursor and select the blocks.

Then press COPY or CUT,

place the cursor where you want to copy or move the selected text. The new position may be in the same program or in another program.

Then press PASTE.

14.5 Program correction

Compare your program to the one in the folder FINISHED_EXERCISES named TEST_14_01.

15. Tool Radius Compensation (1h)
(Theory: 0.5h, Practice: 0.5h)

15.1 Introduction

When offsetting a tool, the distance between the tip of the tool and the characteristic point of the slide on all the axes on which the slide moves (in this case X and Z), is entered into the specific geometry page (paragraph 6.2.2).

Whether these values are obtained by touching the workpiece or by measuring them outside of the machine, the point defined on the X-axis does not correspond to the point defined on the Z-axis.

This is due to the presence of the insert radius.

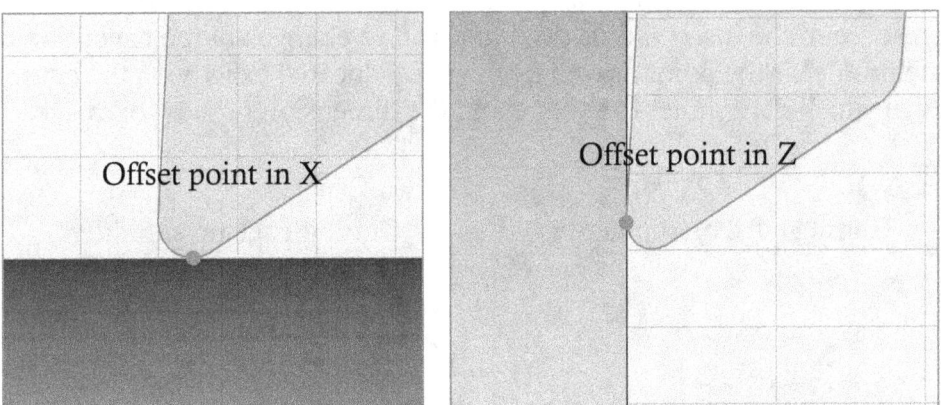

Fig. 112. Offset points on the X-axis and Z-axis with tool radius

The distance between the two points increases when the dimension of the insert radius increases.

The offset values in X and Z define the coordinates of the point used by the NC in order to execute the programmed path. This is on the cutting edge without considering the presence of the insert radius, as if the tool had a sharp edge (see fig. 113).

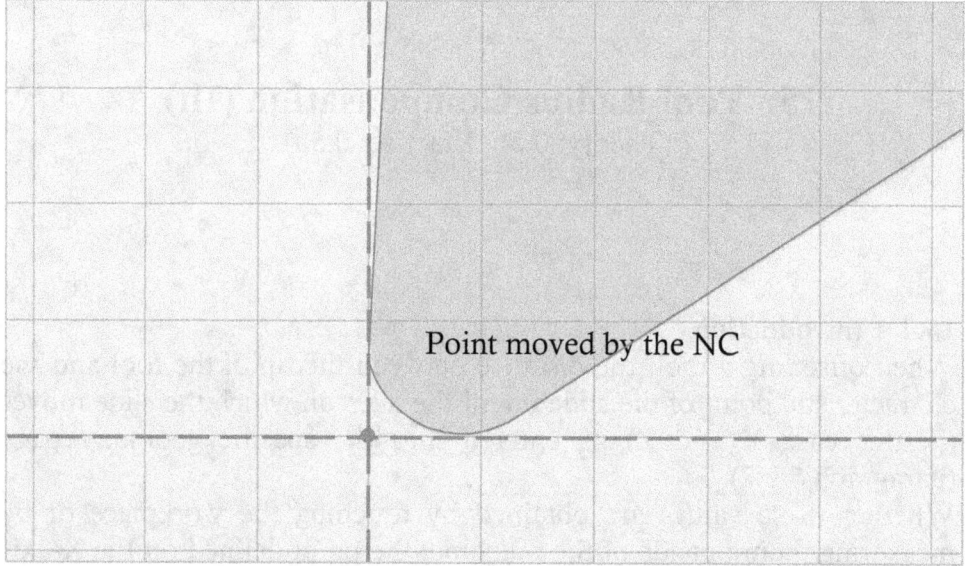

Fig. 113. Point moved by the NC after the tool offset

When executing cylindrical turnings and vertical shoulderings, the presence of the insert radius does not lead to changes in the execution of the profile, as the point moved by the NC is located exactly on the cutting edge which determines the shape and the dimension of the workpiece.

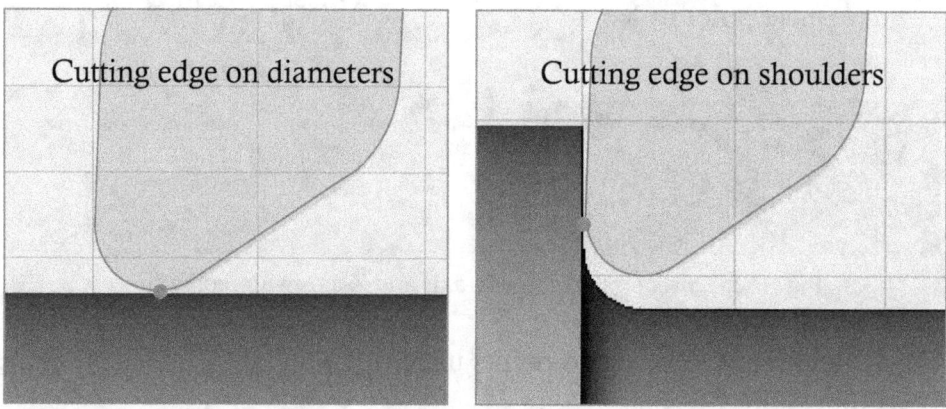

Fig. 114. Absence of changes due to the tool radius on diameters and shoulders

The presence of the tool radius means errors in the description of the profile during the turning of conical workpieces and in the execution of circular interpolations.

When the tool travels along conical profiles, the point moved by the NC does not correspond to the edge of the tool cutting the material. This leads to an error in the dimensions of the described profile.
The inclination angle does not change and keeps the created conical shape geometrically correct.

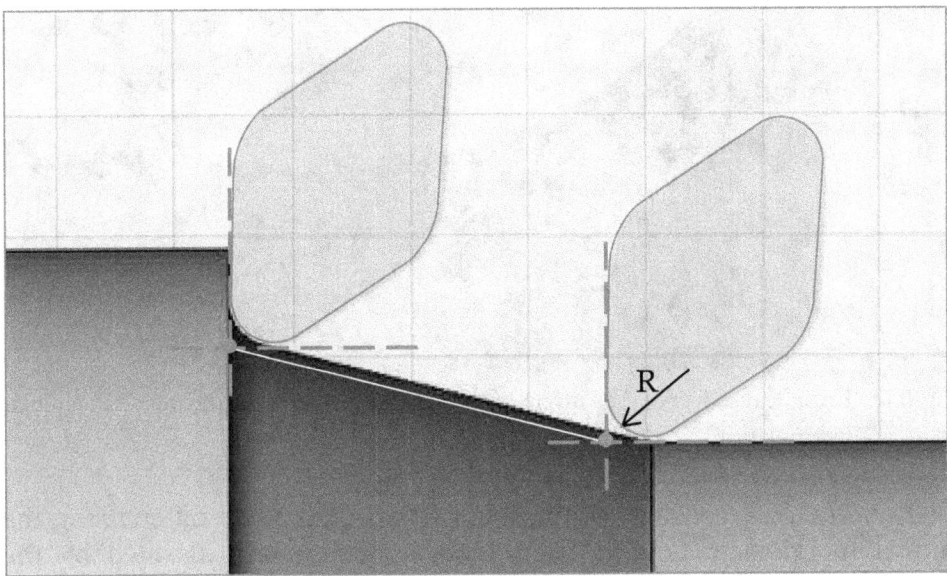

Fig. 115. Dimensional error caused by the insert radius during the execution of conical turnings

As you can see in the figure, the programmed profile, represented by the white line, does not correspond the profile created by the tool.

Also in the execution of circular interpolations, the described profile does not correspond to the programmed profile.
In the following figure you can see how different the programmed profile is from the one actually created by the tool.

Fig. 116. Error caused by the insert radius during the execution of a circular interpolation

The automatic correction of the tool path is performed by enabling the modal functions G42 and G41; these functions are disabled by the function G40.

The information necessary for the NC for the automatic correction of the tool path are:
- dimension of the tool radius
- position of the radius with respect to the zero point

This information is entered in the geometry page during the graphic description of the tool (s. fig. 67 and 68 in chapter 7).

In all the NCs that do not have a graphic description of the tool, the orientation of the zero point compared to the radius (also called the quadrant of the tool) is defined according to an ISO Standard code shown in the following figure.

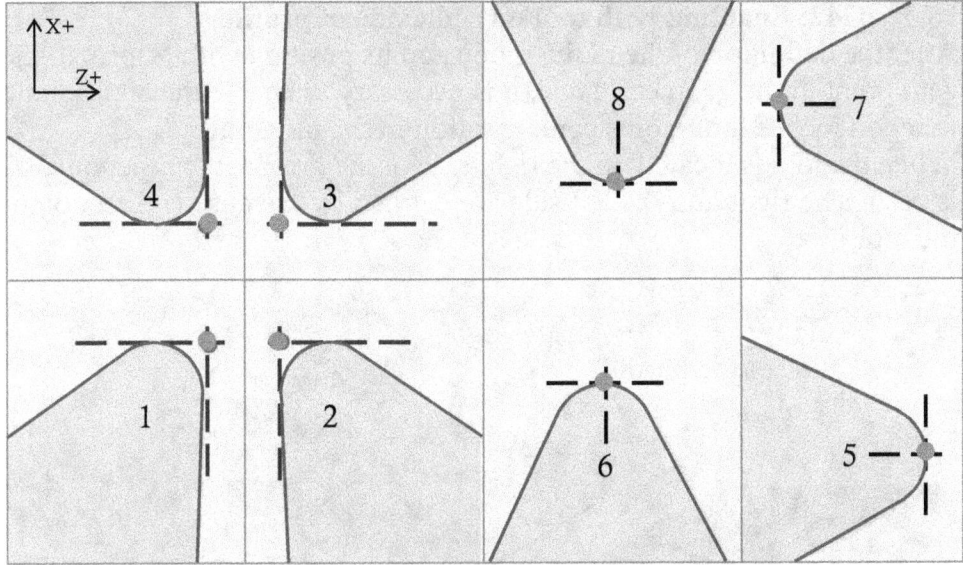

Fig. 117. Quadrant code defining the radius position with respect to the zero point

When the zero point is at the center of the radius which needs to be compensated (as with mills), the quadrant code defining the radius must be 0 or 9.

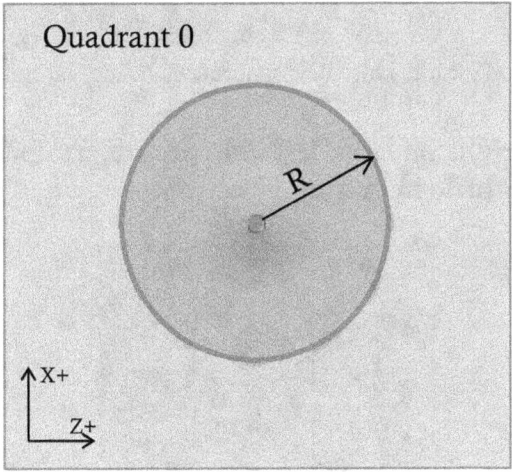

Fig. 118. Quadrant code 0 or 9 for tools offset at the radius center

15.2 G42: Enabling with tool on right side of profile

After the definition of the radius value and its position with respect to the zero point in the geometry table, it is necessary to enable the appropriate function for the tool radius compensation in the program.

When the tool is located on the right side of the profile, the function G42 is used. The right and the left side are defined by the cutting direction of the tool.

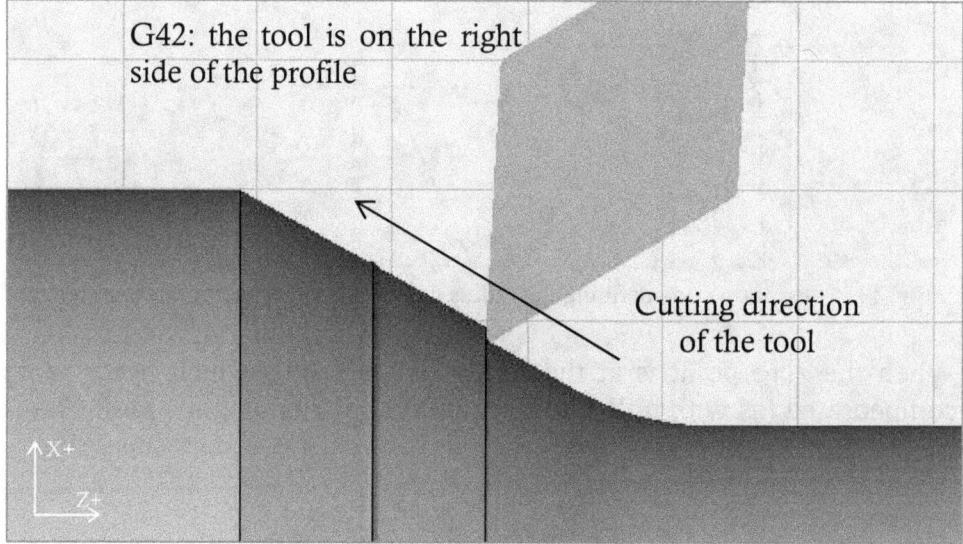

Fig. 119. G42: radius tool 0.2, quadrant 3, to the right of the profile

Attention! The right and the left side are determined as if you were walking on the profile in cutting direction.

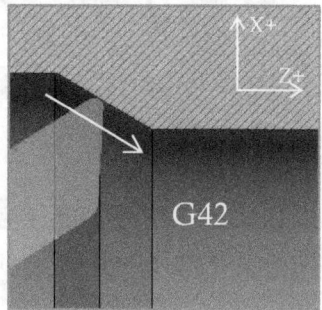

Fig. 120. G42: radius tool 0.8, quadrant 1, to the right of the profile

15.3 G41: Enabling with tool on left side of profile

With the tool on the left side of the program, the function G41 is programmed.

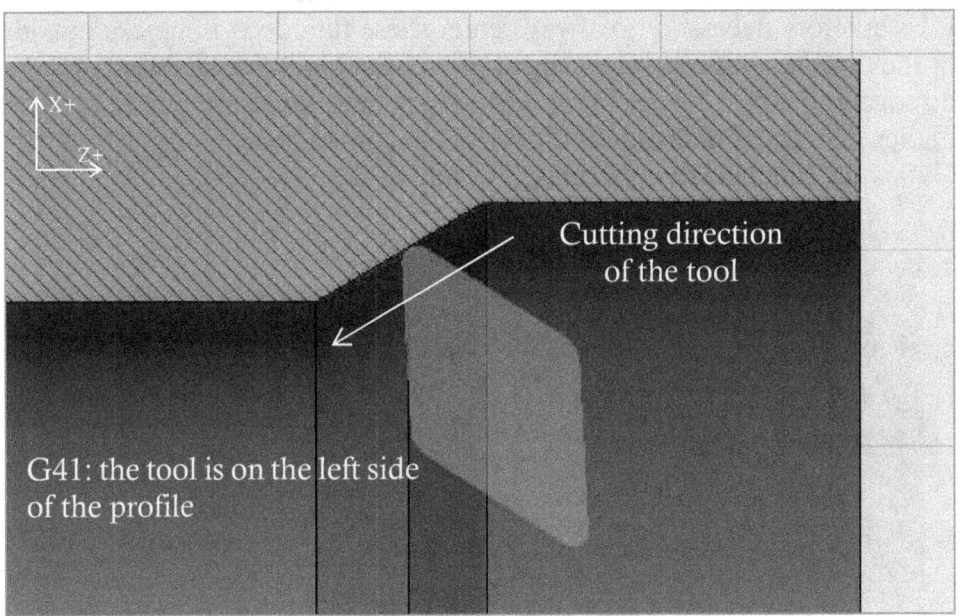

Fig. 121. G41: radius tool 0.8, quadrant 2, to the left of the profile

Make sure not to associate the functions G42 or G41 with external or internal profiles, as **the only correct assessment is to recognize the position of the tool to the right or to the left of the profile with respect to the cutting direction.**

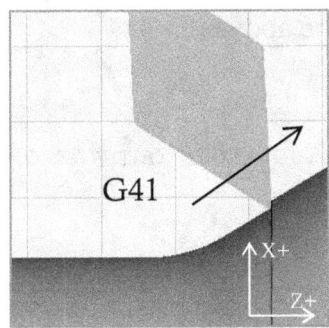

Fig. 122. G41: radius tool 0.2, quadrant 4, to the left of the profile

15.4 Enabling and disabling with G40

In the blocks with the enabling and disabling functions for the tool radius compensation, the NC corrects the programmed path in order to be prepared to execute the profile in a correct manner.

It is therefore necessary to always enter these functions in blocks outside of the profile, as for example the approaching block for enabling and the disengaging block for disabling. Make sure that the programmed offset is higher than the dimension of the radius to compensate.

G40 is the function that disables G41 and G42.

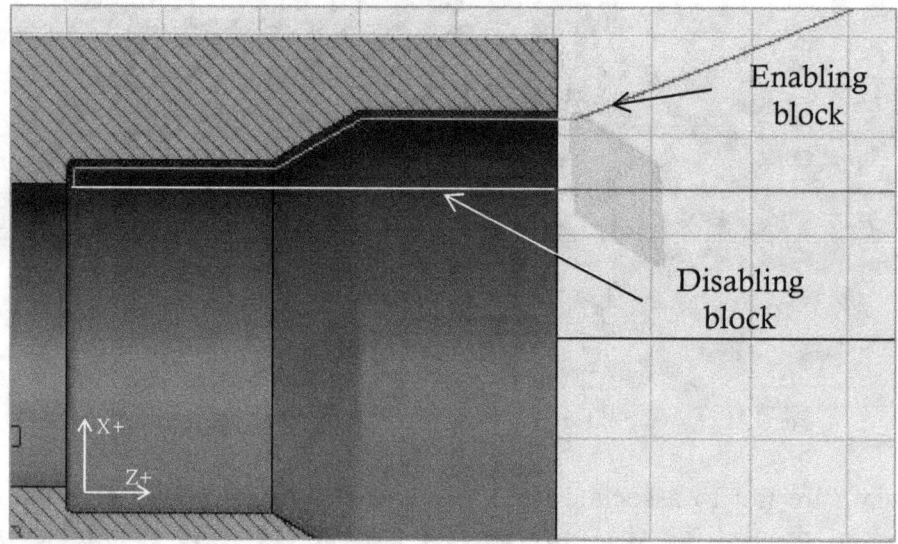

Fig. 123. Enabling and disabling block outside of the profile

Below you will find a programming example for the definition of the profile shown in the figure above.

```
T12 D1
G95 S1800 M4
G0 X45 Z2 G41 ; ENABLING BLOCK OUTSIDE OF PROFILE
G1 Z-20 F0.1
G1 X35 ANG=210
G1 Z-50
G1 X28
G0 Z2 G40 ; DISABLING BLOCK OUTSIDE OF PROFILE
G0 Z200 X200
```

15.5 Practical exercise

15.5.1 Program analysis
Enter the missing data into the program below.
The arrow (→) before the block number tells you where to enter the missing value.

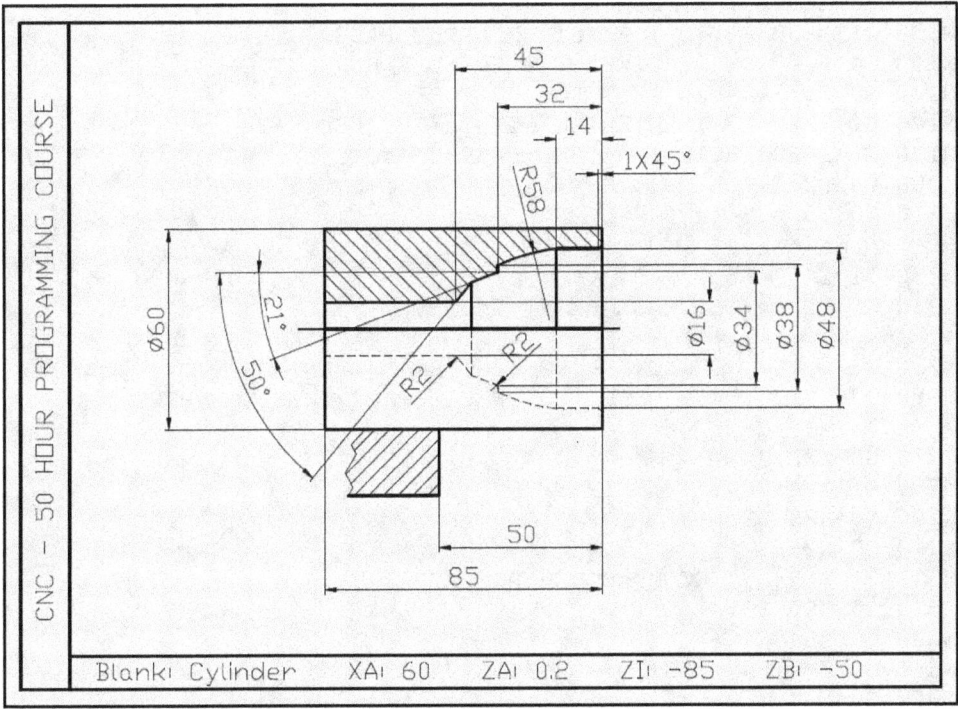

Fig. 124. Drawing of the part to create

```
; blank part dimensions:
; XA = 60 bar diameter
; ZA = 0.2 machining allowance on front face
; ZI = -85 length of finished part
; ZB = -50 extension from jaws
N10 WORKPIECE(,,,"CYLINDER",0,0.2,-85,-50,60)

N20 G18 G54 G90
N30 G0 X400 Z500
N40 M8
N50 SETMS(1)

N60 LIMS=3000 ; LIMITATION TO 3000 REV/MIN
N70 T1 D1 ; TURNING TOOL FOR EXTERNAL PARTS
```

```
N80 G96 S100 M4 ; ENABLING OF CONSTANT CUTTING SPEED

N90 G0 X62 Z0 ; APPROACH
→ N100 G1 X............ F0.18 ; FACING
N110 G0 X200 Z200 ; DISENGAGEMENT

N120 T11 D1 ; RIGHT AXIAL DRILL DIAMETER 16 MM
N130 G95 S1100 M3 ; ENABLING OF FIXED NUMBER OF REVOLUTIONS
N140 G0 X0 Z2 ; APPROACH
N150 G1 Z-30 F0.12 ; FIRST DRILLING PASS
N160 G4 S2 ; DWELL TIME OF 2 SPINDLE REVOLUTIONS
N170 G0 Z5 ; RAPID EXIT FOR CHIP REMOVAL
N180 G1 Z-29 F2 ; ENTERING WITH HIGH FEED RATE
N190 G1 Z-60 F0.12 ; SECOND PASS UP TO Z-60
N200 G4 S2
N210 G0 Z5
→ N220 G1 Z............ F2
N230 G1 Z-90 F0.12
N240 G0 Z200 ; DISENGAGEMENT IN Z

N250 T12 D1 ; BORING BAR FOR INTERNAL TURNING
N260 G96 S120 M4 ; ENABLING OF CONSTANT CUTTING SPEED

N270 G0 X22 Z2
N280 G1 Z-41 F0.14 ; FIRST ROUGHING PASS
N290 G0 X20 Z2
N300 G0 X28
N310 G1 Z-34 ; SECOND ROUGHING PASS
N320 G0 X26 Z2
N330 G0 X34
N340 G1 Z-31.8 ; THIRD ROUGHING PASS
N350 G0 X32 Z2
N360 G0 X40
N370 G1 Z-28 ; FOURTH ROUGHING PASS
N380 G0 X38 Z2
N390 G0 X46
N400 G1 Z-16 ; FIFTH ROUGHING PASS
N410 G0 X44 Z5

;BEGINNING OF FINISHING WITH SAME TOOL
N420 G96 S150 M4 ; ENABLING OF CONSTANT CUTTING SPEED

N430 G0 X50 Z5
N440 G0 Z2 G41
N450 G1 Z0 F0.1
N460 G1 Z-1 ANG=225
→ N470 G1 Z............
→ N480 G3 X............ Z-32 CR=58
```

```
→ N490 G1 X34 RND=............
  N500 G1 ANG=201
→ N510 G1 X............ Z-45 ANG=230 RND=2
  N520 G1 Z-48
  N530 G1 X15
  N540 G0 Z5 G40
  N550 G0 X200 Z200
  N560 M30
```

Now open the program EX_15_01 in the folder 01_EXERCISES and enter the values.

Before starting the graphic simulation create both of the tools used in this cycle (paragraph 7.5):
- in position 11, an axial drill diameter 16 mm
- in position 12, a boring bar.

Use the names and data shown in the figure below.

11	⚙	AX. DRILL D.16	1	1	100.000	120.000	16.000		118.0		
12	⚙	ROUGH. BORING-BAR	1	1	86.000	92.000	0.400 ←		93.0	55	8.0

Fig. 125. Data of the new tools to create for the execution of the cycle

If there are problems with the creation of the new tools, you can review the practical exercise in paragraph 7.5, or proceed by loading the tool file named TOOL_LIST from the folder 01_EXERCISES according to the procedures laid down in paragraph 3.3.

Start the graphic simulation in single block mode and display the HALF CUT VIEW of the workpiece, analyze the program and change it where necessary.

Compare your program to the one in the folder FINISHED_EXERCISES named EX_15_01.

15.5.2 Test of concept comprehension

The following exercise proposes different combinations of tool radius compensation enabling function and radius quadrant code.
Put a cross by the only correct combination among the given answers.

1)

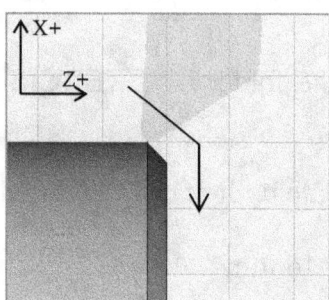

a) G42, quadrant 3

b) G41, quadrant 3

c) G41, quadrant 4

2)

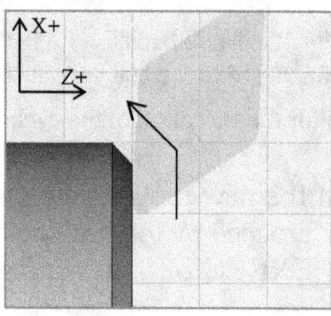

a) G41, quadrant 1

b) G42, quadrant 3

c) G42, quadrant 2

3)

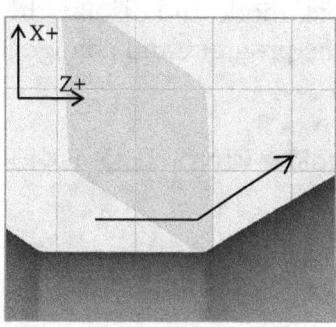

a) G42, quadrant 4

b) G42, quadrant 2

c) G41, quadrant 4

4)

a) G41, quadrant 2

b) G42, quadrant 2

c) G41, quadrant 3

5)

a) G42, quadrant 2

b) G41, quadrant 2

c) G42, quadrant 1

6)

a) G41, quadrant 6

b) G42, quadrant 3

c) G42, quadrant 8

The correct solutions can be found in the program ANS_15_01 in the folder CHAP_15.

15.6 Reloading of complete tool list
Before reading the next chapter, reload the complete tool list contained in the folder 01_EXERCISES.

16. Three-Axis Mill: Programming (2h)
(Theory: 2h)

16.1 Introduction
The ISO functions applied to the lathe up until now are the same functions which allow for the programming of a 3-axis mill.

In a lathe, the main working plane is the X-Z plane defined by the function G18.

In a 3-axis mill, the main working plane is the X-Y plane defined by the function G17. On this plane, the tool, rotated by the spindle, moves to execute the profile described in the program, while the position of the tool on the Z-axis determines the execution depth of the tooling operation.

16.2 Layout of the axes in a mill
The scheme for the layout of the axes is the one shown in chapter 4.

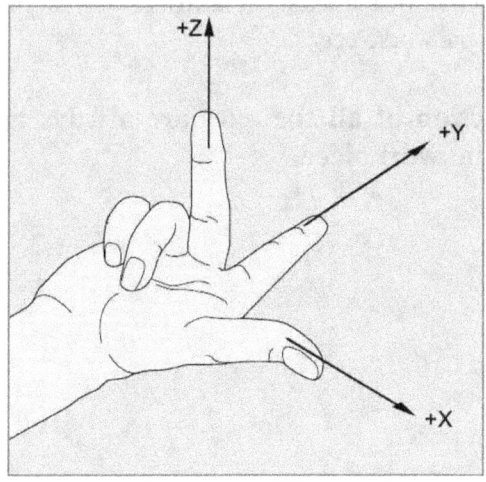

Fig. 126. The same right hand rule applies both to the lathe and to the mill.

The machine under examination is a mill with three axes: X, Y, and Z. The axes X and Y are applied to the machine table while the Z-axis is applied to the tool carrying slide.

Fig. 127. Positive direction of the axes: the arrows show the movements of the tool compared to the workpiece

The positive direction of all the axes are always applied to the tool which moves on the workpiece.

16.3 The X-axis, the Y-axis and the Z-axis

In a mill, the axes that determine the main working plane are the X-axis and the Y-axis. In the X-Y plane (G17), the profile of the workpiece is normally programmed. The Z-axis is the axis that determines the machining depth. The mill is vertical when the axis around which the tool rotates (Z-axis) is positioned vertically; it is horizontal when the Z-axis is oriented horizontally.

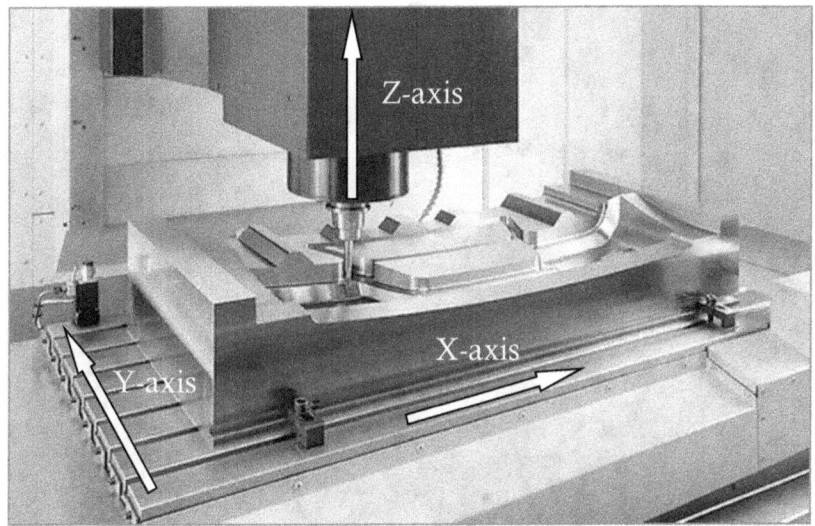

Fig. 128. Vertical mill

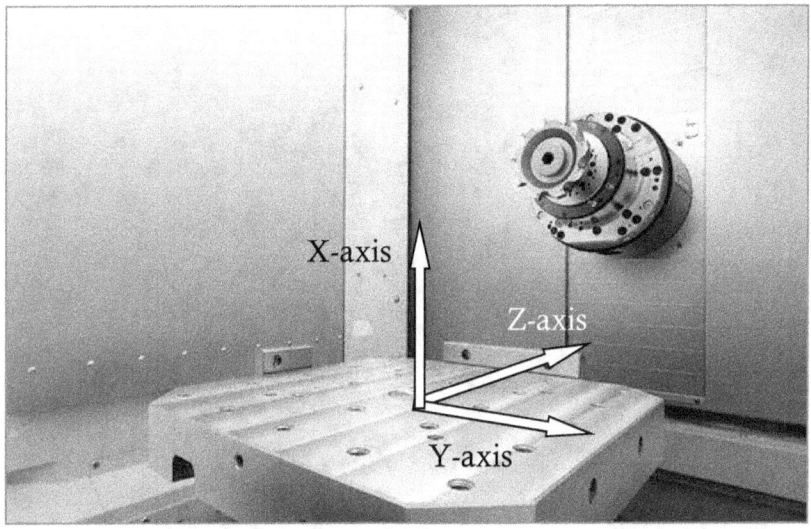

Fig. 129. Horizontal mill

16.4 C- and B-axis in machining centers

When equipped with additional rotating axes, mills are called machining centers. If the rotating axis rotates around Z, as already seen in the lathe, it is called the C-axis. If the rotating axis rotates around Y, it is called the B-axis. The simultaneous presence of two rotating axes makes the presence of the third rotating axis unnecessary. For this reason the most complete configuration of a machining center is with five axes and not with six axes.

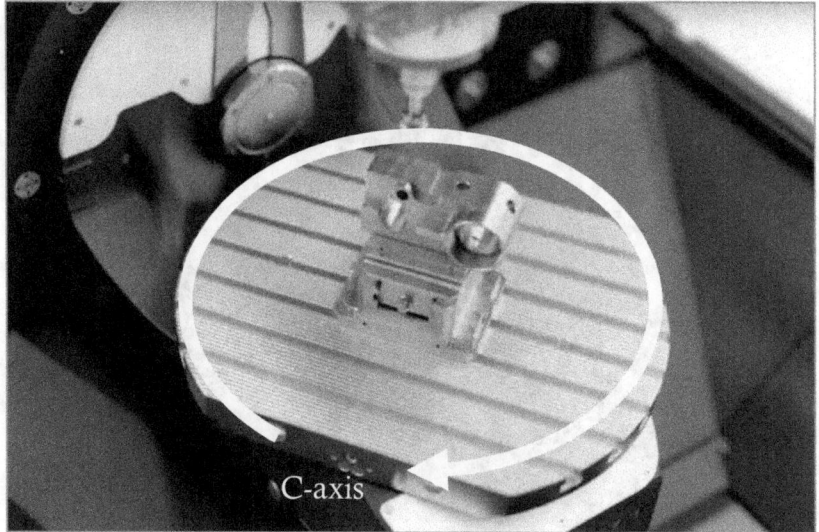

Fig. 130. C-axis in a machining center

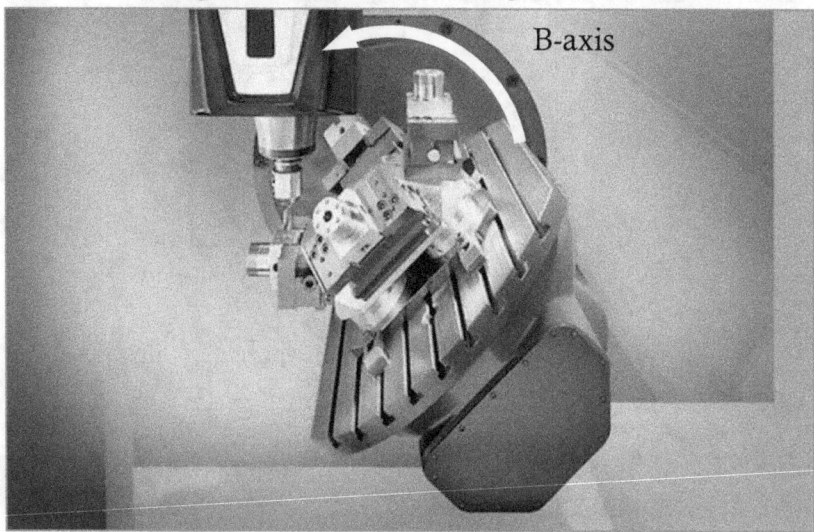

Fig. 131. B-axis in a machining center

16.5 Five-axis interpolation

We have already seen that interpolation is the coordinated movement of one or more axes according to a precise geometric logic performed with specific speed.

The geometric logic of a line is described on a plane defined by two or three linear axes.

The geometric logic of a circle arc is described on a plane defined by two linear axes (or one linear and one rotating axis).

The geometric logic of a helix is defined in the space in a three-dimensional system defined by three linear axes (or two linear axes and one rotating axis).

In a reference system consisting of more than three axes, there is no element that can be defined according to a geometric rule.

That is why, when we speak of five-axis interpolation, we are talking about simple point-to-point programming; i.e. each point of the profile is reached by a straight line whose movement causes the remaining two rotating axes to reach a different angular position at the same time.

This type of programming is always done through CAM software that analyzes the three-dimensional drawing of the workpiece to be produced and generate the profile to be machined by breaking it down into multiple sections and rotating the remaining rotating axes to orient the tool in order to keep the cutting angle constant and not cause collisions with the workpiece.

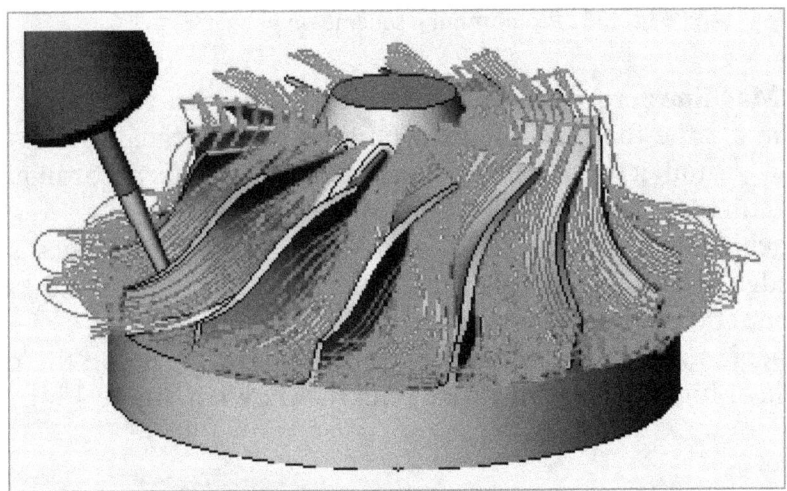

Fig. 132. CAM-generated profile for five-axis machining center

16.6 Programming scheme

The following figure shows the programming scheme applied to plane G17. It has to be used for the assessment of the positive direction of the axes, in order to determine the clockwise and counterclockwise rotation of the circle arcs included in the profile of the workpiece, to define the right and left position of the mill with regard to the cutting direction and to assess the angle to be programmed in the event of inclined lines.

Standard ISO 841 defines a coordinate system of axes where the positive direction is always referred to the movement of the tool.

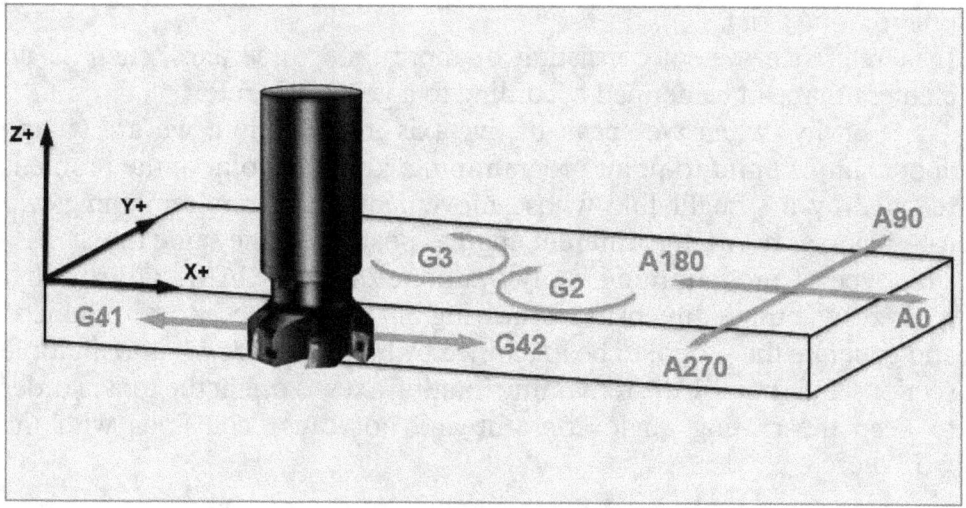

Fig. 133. Programming scheme on plane G17

16.7 Machine zero point and definition of the part zero point

While in a lathe the machine zero point is often located on the spindle nose, in a mill it varies from machine to machine according to the manufacturer's choices.

The machine zero point on the X- and Y-axes in some machines is placed on an edge of the table, in other machines it is at the center of the table at the intersection of the diagonals.

The machine zero point on the Z-axis can be positioned on the plane of the table or high up close to stroke limit (as shown in figure 134).

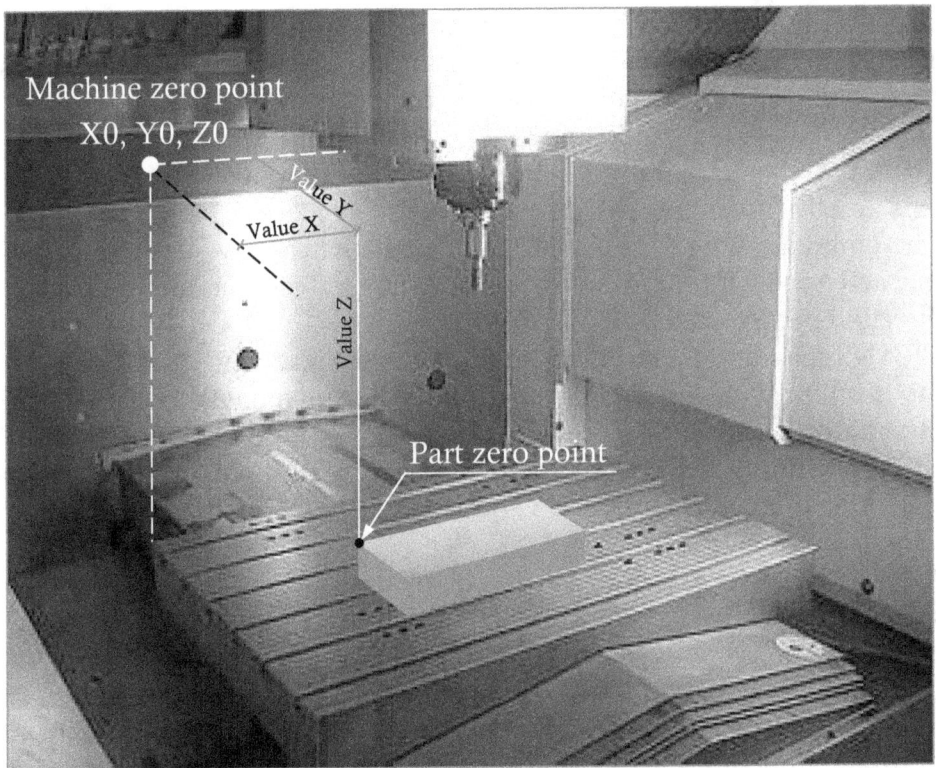

Fig. 134. Position of the machine zero point and values to enter into the zero offset function for the definition of the part zero point

The part zero point is defined by moving the machine zero point on the three axes to the point chosen by the operator. The functions to be used are those for the zero offset (G54, ..., G57) seen in chapter 6.

The values to enter correspond to the distance between the machine zero point and the part zero point as shown in the previous figure.

On the basis of the scheme showing the positive direction of the axes (figure 127), it can be deduced that in this case the value of Z is negative, the value of X is positive and the value of Y is negative.
The manufacturer's manual gives the position of the machine zero point.

16.8 TRANS/ATRANS: incremental shift of the part zero point

The TRANS function allows for the shifting by the program of the part zero point by incrementing the values entered into the absolute zero offset functions: G54, G55, G56 and G57.

The TRANS command must be followed by the letter referring to the axis and by the incremental offset value to execute.

The programming of 'TRANS Y50' means that one wants to increment the active absolute zero offset (e.g. G54) by 50 mm in the positive direction of the Y-axis.

TRANS may be programmed on all the linear axes (X, Y, Z) present in the machine. The ATRANS function further increments the zero offset programmed with the TRANS function.

In a machining center, these functions can be used to replicate the execution of a program part at various points of the workpiece (see figure 135-2).

Another option is to mount various blank parts on the machine table and shift the complete execution of the program from one blank part to the other.

In a lathe these functions are more commonly used when it is decided to define the part zero point within the program. It is for example possible to use the function G54 to shift the machine zero point to a fixed point (e.g. the face of the jaws) to then program the TRANS function, followed by the letter 'Z' associated with the extension of the workpiece face from the jaws (see figure 135-1).

To cancel the function program TRANS X0 Y0 Z0.

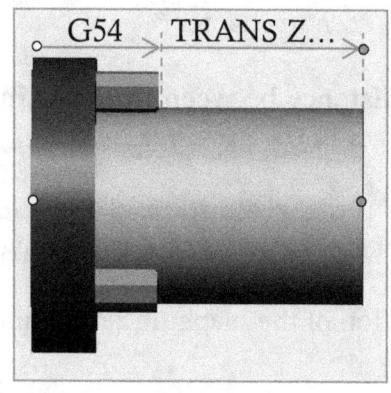 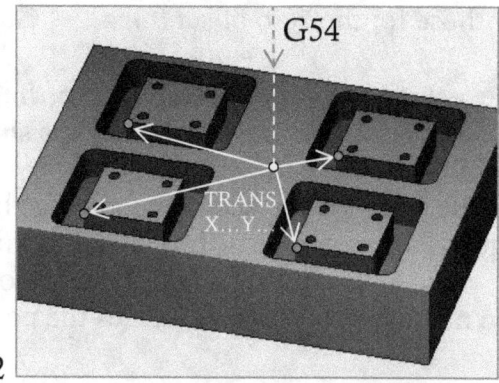

Fig. 135. Use of TRANS: 1: in a lathe; 2: in a machining center

16.9 Position of the point controlled by the NC and tool geometry

The point controlled by the NC is always placed by the manufacturer at the spindle's center of rotation on the tool holder attachment plane (fig. 136-1). The tool attachments are normally standardized and have a conical shape of varying dimension according to the maximum dimension of the tools which can be mounted to the machine.

The rotating center of all the tools used in milling machines is concentric to the point moved by the NC. This means that the geometry values on the X- and Y-axes are always equal to zero, while the zero offset value on the Z-axis corresponds to the distance between the tip of the tool and the attachment plane of the tool holder (fig. 136-2).

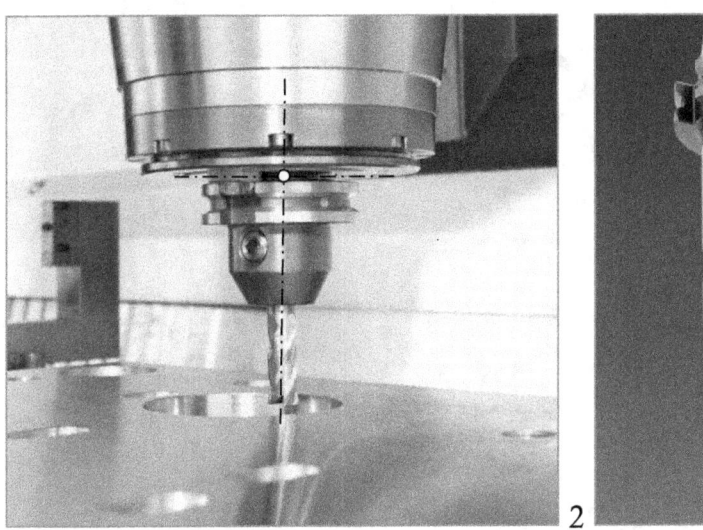
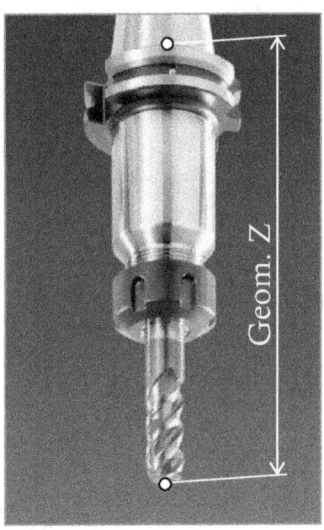

Fig. 136. 1:Point moved by the NC; 2:Offset value of a mill on the Z-axis

To obtain the offset value in Z by touching the workpiece:
- you program the activation of the part zero point in MDA by using the zero offset function set in the program (G54),
- you touch the workpiece at a known value with respect to the part zero point,
- you enter this value and you activate the automatic calculation.

Another method is to measure the tool geometry outside of the machine either using an altimeter or by means of specific external measuring systems as the one shown in figure 137.

The systems for external presetting have a tool attachment which is identical to the one present in the machine. Before proceeding to the measuring of the tool length, the tool holder is offset by making the measured zero coincide with the point moved by the NC (in this case on the face of the attachment cone); then the tool is fixed to the support in order to measure its length and, if necessary, other characteristic elements like the diameter or the length of the cutting edge.
A camera with video helps the operator to define exactly the measuring point.

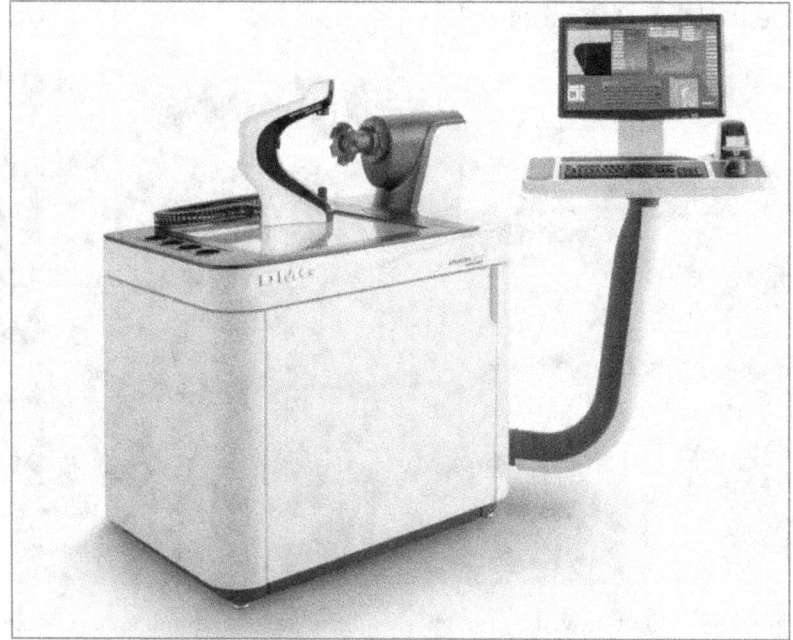

Fig. 137. Measuring system outside the machine

The diameter of the mill (inserted in the geometry page) will be considered by the machine by programming of the tool radius compensation functions.
Contrary to lathes, in mills the concept of tool quadrant code does not exist, as all the tools used in those machines are offset at the center and are therefore always defined by quadrant zero. **To review the tool radius compensation functions and the quadrant codes, please refer to chapter 15 and to the programming examples in the next chapters.**

16.10 Setting of tool rotation and feed rate

In chapter 8, the functions for the activation of the spindle rotation in a lathe have been described. In a mill, this is much more simple.

In the lathe, the setting of the spindle rotation with fixed number of revolutions or with constant cutting speed was examined; in the mill, the only alternative is a fixed number of revolutions calculated on the basis of the tool diameter used.

The number of revolutions of the mill is calculated using the formula shown in figure 75.

In the case of a mill with a diameter of 32 mm, which works at a cutting speed of 100 m/min., the number of revolutions to be programmed corresponds to:

$$n = (1000 \times 100) / (32 \times 3.14) = 995 \text{ rev./min.}$$

In chapter 9, the functions for the setting of the feed rate were examined. In this case the functions G95, G94 and all the concepts associated with them are applicable to the mill exactly as already seen for the lathe.

17. Practical Milling Exercise (3h)
(Practice: 3h)

17.1 Introduction
Having clarified the use of the ISO functions, we will now see a complete programming example for a part created on a 3-axis mill.
The drawing shown in figure 147 represents the part to be created.

17.2 Creation of a three-axis mill (X, Y, Z)
Before proceeding, it is necessary to create the milling machine to be used in this chapter in the training and simulation software.

Start SinuTrain and click on "Use Template".

Now choose the machine type, change the name of the mill, describe its basic features, set the size of the window which reproduces the machine video and the language you want to use. Enter the following information:

Template	DEMO-Milling Machine
Name:	MILL: Programming Course
Description:	SP1-spindle (main spindle)
	X-axis (linear geometry axis)
	Y-axis (linear geometry axis)
	Z-axis (linear geometry axis)
Resolution:	640x480 (or other resolution that best fits your screen)
Language:	English - English

Push CREATE.

The machine has been created and is now displayed on the starting page of the program.

170

Click on the newly created icon to start the mill.

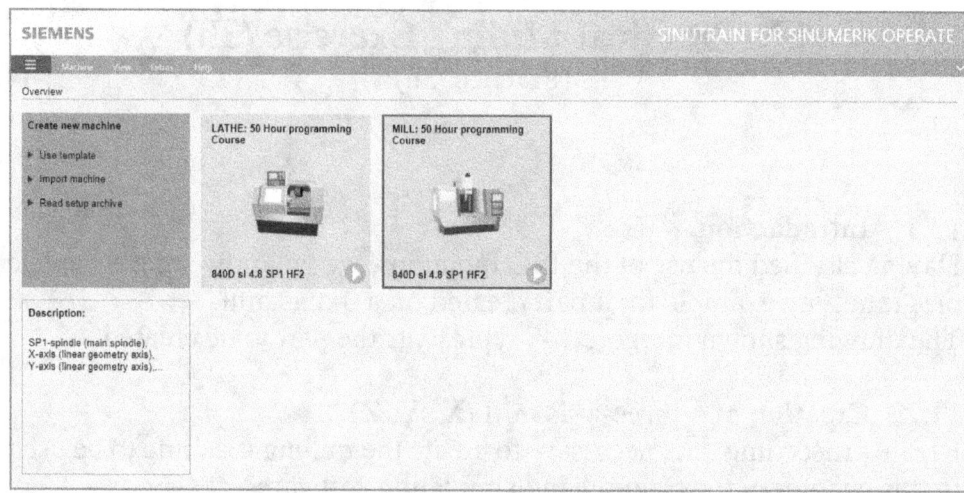

Fig. 138. Start-up of the mill in the training software

17.3 Download of the programs and import into SinuTrain

Open the website cncwebschool.com and access the TOOLS area to download the folder M3_PROG containing the programs related to the mill.

Select the compressed folder you just downloaded with the pointer, push the right button of your mouse and select: *Extract all*.

Now import the programs into the training software.

Copy the folder M3_PROG onto an empty USB stick.

On the control panel, click PROGRAM MANAGER.
After selecting USB from the horizontal softkeys, the content of the USB memory is shown.

Select the folder M3_PROG with the arrows and **push the yellow INPUT button to open it.**

Select all the folders with the softkey SELECT.
Press the vertical softkey COPY.
Push NC from the horizontal softkeys.
Move down with the arrows until you have selected the folder WORKPIECES, then press PASTE from the vertical softkeys.
Open the WORKPIECE folder with INPUT, open the folder CHAP_17 with INPUT and, also with INPUT, open the test program PRG_0.

17.4 Direct selection of the tools in the program

The creation of new tools and the export and import of tooling data follow the same rules as seen in chapter 7.
For this exercise it is not necessary to set any new tools as the program uses the tools which are already present in the machine.

It is furthermore possible to use a new method to select the tools in the program: instead of programming the position T and the geometry D, the tool to be used is chosen directly from the magazine.

Position the cursor on the block where you want to call the tool and press the vertical button SELECT TOOL.

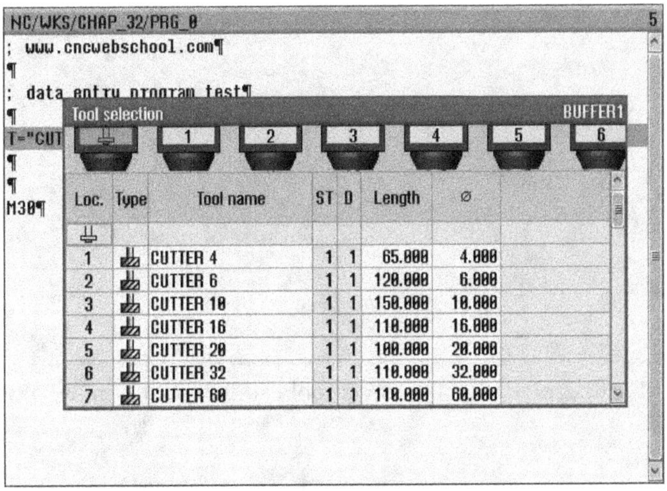

Fig. 139. Page for the selection of the tools directly from the magazine

A list of available tools is shown.
With the arrows, highlight the tool to be called and confirm your selection with OK. The name of the tool is inserted in the program.

As will be seen in the example program, **the tool selection needs to be completed by adding the function D1 to the same block to retrieve the offset values and the function M6 to enable the tool changing procedure**.
The block must be programmed as follows:

T="CUTTER 16" D1 M6

For the selection of the tool offset values see paragraph 7.3, for the function M6 see figure 45 and paragraph 7.2.

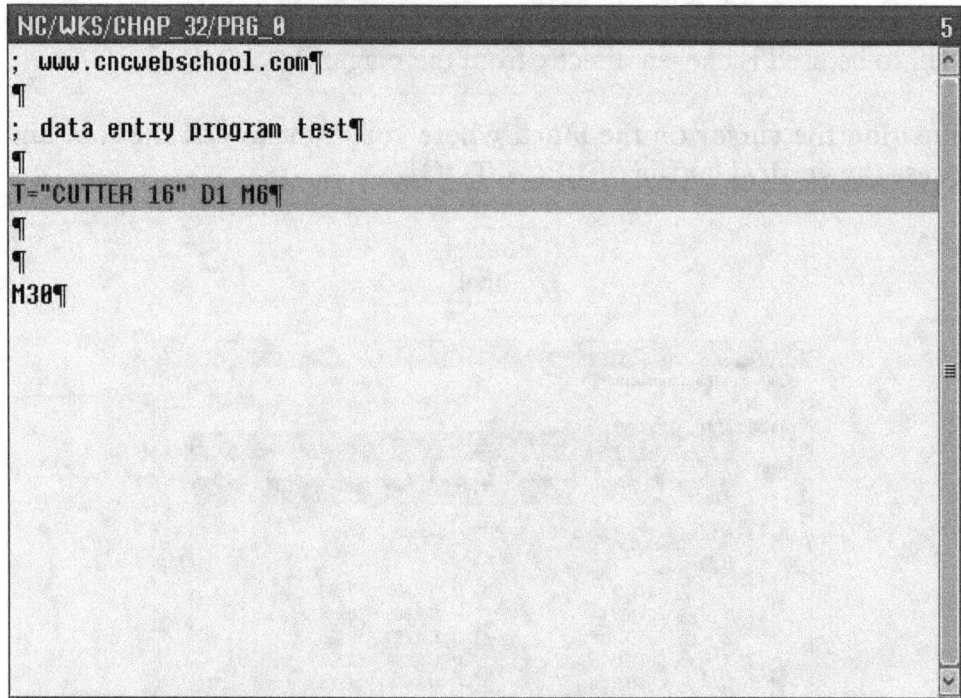

Fig. 140. Completion of the direct tool call instruction in the program

17.5 Graphic definition of the blank part

Now close the program PRG_0.
Open the program PRG_17_01 and open, at block N10, the dialog box for the insertion of the blank part data.

N10 WORKPIECE(,,,"RECTANGLE",0,0,-32,-150,160,120)

By means of the graphic definition of the blank part the following elements are defined:
- shape of the blank part,
- position on X and Y of the part zero point,
- position on Z of the part zero point,
- dimensions of the blank part.

Below is the explanation of the parameters based on the different options that may be selected.

The selection CYLINDER forces the position of the part zero point in X and Y onto the main axis of the cylinder.

Blank part:	Cylinder
XA:	Cylinder diameter.
ZA:	Position of the upper face of the workpiece referring to the part zero point.
ZI - absolute: - incremental:	Distance from the lower face of the workpiece: **referring to the part zero point.** **referring to the upper face.**

Fig. 141. Description of the blank part dimensions: CYLINDER

Below are some examples applied to a 32 mm high blank part which help understand how the graphic simulation interprets the values associated with ZA and ZI.

If you want to simulate the position of the **part zero point on the upper face of the blank part,** set the following values:

ZA = 0: shows that the upper face of the blank part corresponds to the position of the part zero point.

ZI = -32 (if the value is expressed in **absolute** coordinates) shows that the lower face of the blank part is located at a distance of 32 mm in the negative direction **compared to the part zero point**.

ZI = -32 (if the value is expressed in **incremental** coordinates) shows that the lower face of the blank part is located at a distance of 32 mm in the negative direction **compared to the upper face**.

If ZA equals zero, the position of the part zero point corresponds to the upper face of the blank part; therefore the value ZB which expresses the total height of the blank part (defining the position of its lower face) is equal both when expressed in absolute coordinates and when expressed in incremental coordinates.

If you want to simulate the position of the **part zero point on the machine table,** set the following values:

ZA = 32: shows that the upper face of the blank part is located at a distance of 32 mm in the positive direction of the part zero point.

ZI = 0 (if the value is expressed in **absolute** coordinates) shows that the lower face of the blank part corresponds to the position of the part zero point.

ZI = -32 (if the value is expressed in **incremental** coordinates) shows that the lower face of the blank part is located at a distance of 32 mm in the negative direction compared to the upper face.

The selection PIPE forces the position of the part zero point in X and Y onto its main axis.

Blank part:	Pipe
XA:	External diameter of the pipe.
XI:	Internal diameter of the pipe.

Fig. 142. Description of the blank part dimensions: PIPE

The selection BLOCK CENTERED forces the position of the part zero point in X and Z onto the intersection of the rectangle's diagonals.

Blank part:	Block Centered
W:	Side of the rectangle positioned along the Y-axis.
L:	Side of the rectangle positioned along the X-axis.

Fig. 143. Description of the blank part dimensions: BLOCK CENTERED

The selection BLOCK refers the basic position of the part zero point in X and Y to the edge at the lower left side of the rectangle.

Blank part:	Block
X0:	Coordinate X of the edge referring to the part zero point.
Y0:	Coordinate Y of the edge referring to the part zero point.
X1:	Coordinate X of the opposite edge referring to the part zero point (abs.) or to the first edge (incr.).
Y1:	Coordinate Y of the opposite edge referring to the part zero point (abs.) or to the first edge (incr.).

Fig. 144. Description of the workpiece dimensions: BLOCK

In the case of the block, in order to bring the part zero point to the center of the diagonals of a rectangle with 160 mm sides on the X-axis and of 120 mm sides on the Y-axis, set the data as shown in the following figure:

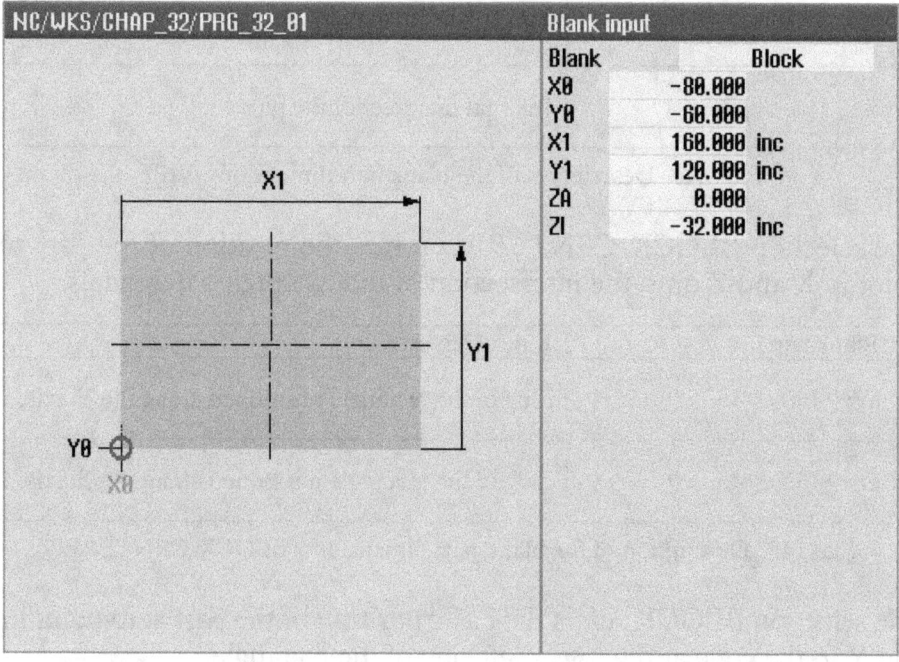

Fig. 145. Description of the blank part dimensions: BLOCK

The selection N CORNER forces the position of the part zero point in X and Y onto the intersection of the polygon's diagonals.

Blank part:	N Corner
N:	Number of edges of the polygon.
SW:	Dimension of the polygon's key (available only for polygons with even number of edges).

Fig. 146. Description of the workpiece dimensions: N CORNER

17.6 Drawing of the part to be created

The part zero point is located on the intersection of the diagonals on the upper face of the workpiece as specified in the drawing.

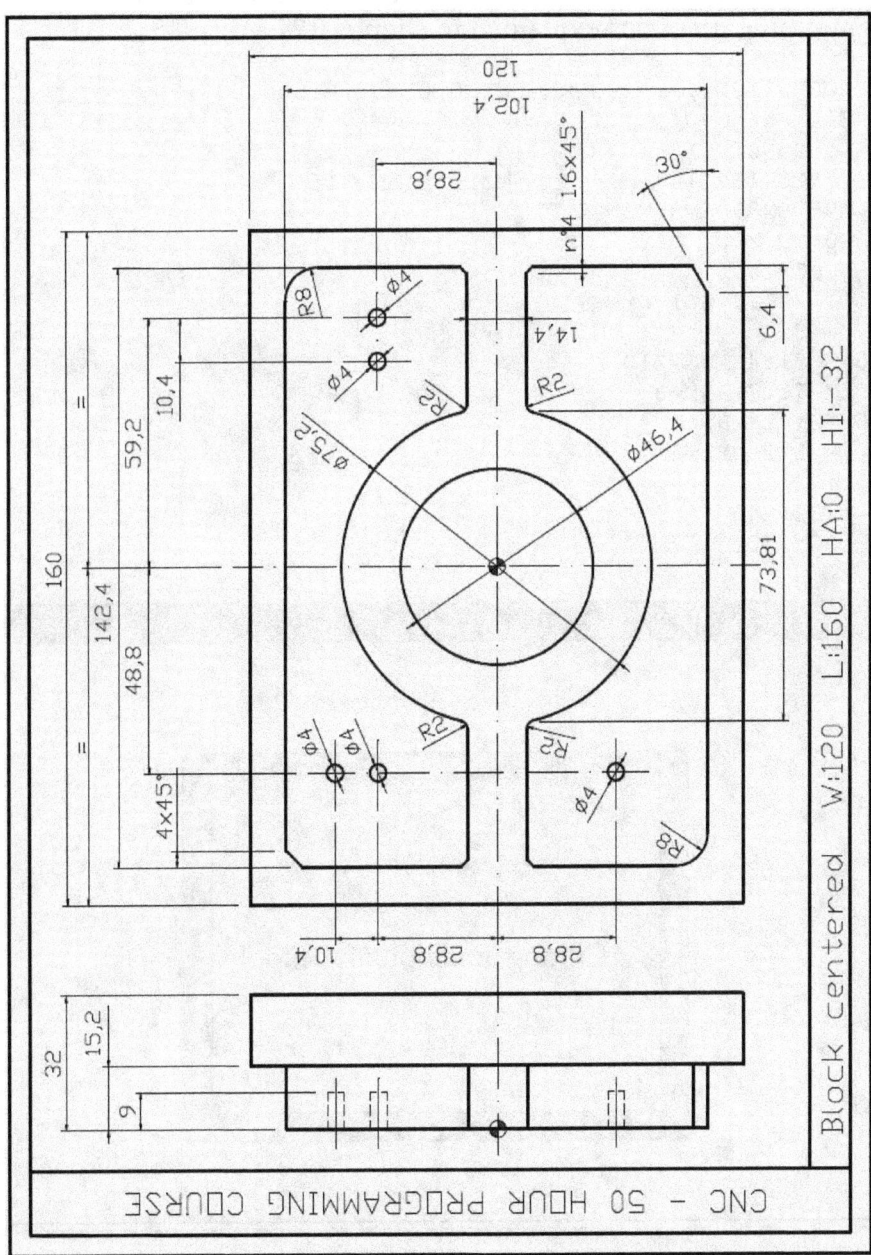

Fig. 147. Drawing of the part to be created

17.7 Program, phase 1: execution of the external profile

Activate the graphic simulation of the program PRG17_01 and associate the tool movements with the blocks programmed hereafter. Chamfers, rounds and angles are defined in the tool path by means of the direct programming functions explained in chapter 12.

```
N10  WORKPIECE(,,,"RECTANGLE",0,0,-32,-150,160,120)
N20  G17 G54 G90
N30  G0 Z500
N40  T="CUTTER 16" D1 M6 ; MILL DIAM. 16
N50  G95 S2800 M3 M8 F0.2
N60  G0 Y0 X90
N70  G0 Z-15.2
N80  G1 X71.2 G41
N90  Y-48
N100 Y-51.4 ANG=210
N110 X-71.2 RND=8
N120 Y51.2 CHF=4
N130 X71.2 RND=8
N140 Y-2
N150 X82 G40
N160 G0 Z500
```

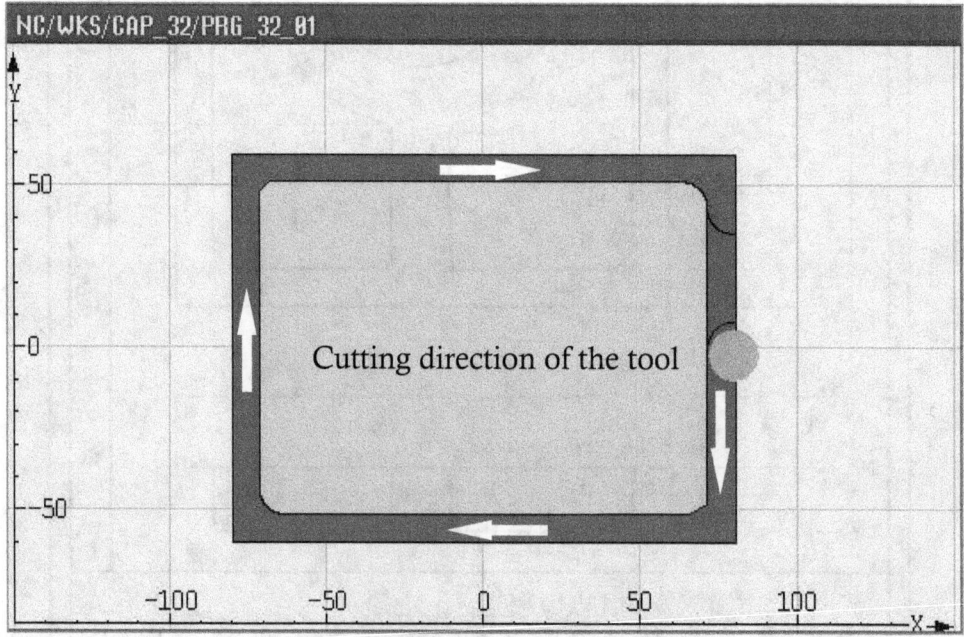

Fig. 148. Creation of the external profile

17.8 Program, phase 2: roughing of the internal profile

The programming of the circular interpolations is carried out according to the instructions in chapter 13.

```
N170 T="CUTTER 10" D1 M6 ; MILL DIAM. 10
N180 G95 S2500 M3 M8 F0.16
N190 G0 Y0 X80
N200 G0 Z-15.2
N210 G1 X30.4
N220 G3 X30.4 Y0 I-30.4
N230 G3 X-30.4 Y0 CR=30.4 F1
N240 G1 X-74

N250 G0 Z500
```

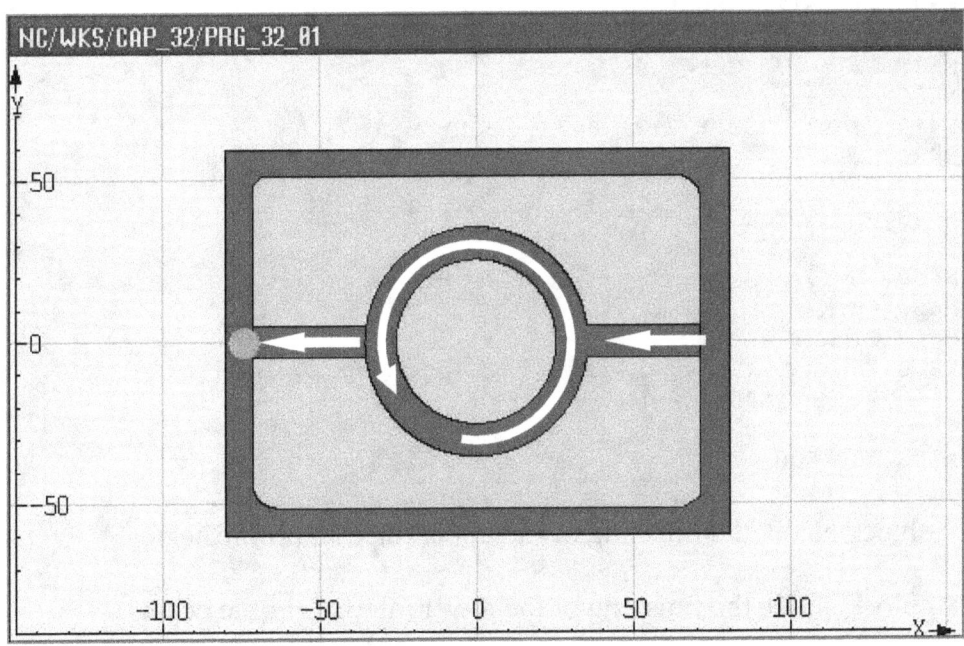

Fig. 149. Roughing of the internal profile

NOTE: the letters I, K and J stand for the coordinates of the radius center referring to the starting point of the arc on the X-, Z-, and Y- axis respectively.

17.9 Program, phase 3: finishing of the internal profile
In this program part the internal profile of the workpiece is finished.

```
N260 T="CUTTER 10" D1 M6 ; MILL DIAM. 10
N270 G95 S3200 M3 M8 F0.16
N280 G0 Y12 X80
N290 G0 Z-15.2
N300 G1 X71.2 G41
N310 G1 Y7.2 CHF=1.6
N320 G1 X36.905 RND=2
N330 G3 X-36.905 Y7.2 CR=37.6 RND=2
N340 G1 X-71.2 CHF=1.6
N350 G1 Y12
N360 G1 X-80 G40

N370 G1 Y-12 X-80 F1
N380 G1 X-71.2 G41 F0.16
N390 G1 Y-7.2 CHF=1.6
N400 G1 X-36.905 RND=2
N410 G3 X36.905 Y-7.2 CR=37.6 RND=2
N420 G1 X71.2 CHF=1.6
N430 G1 Y-12
N440 G1 X80 G40

N450 G0 Y0
N460 G0 X32
N470 G1 X23.2 G41
N480 G2 X23.2 Y0 I=-23.2
N490 G1 X32 G40

N500 G0 Z500
```

At block N260, the finishing of the upper internal profile begins.

At block N370, the finishing of the lower internal profile begins.

At block N450, the finishing of the internal diameter of 46.4 mm begins.

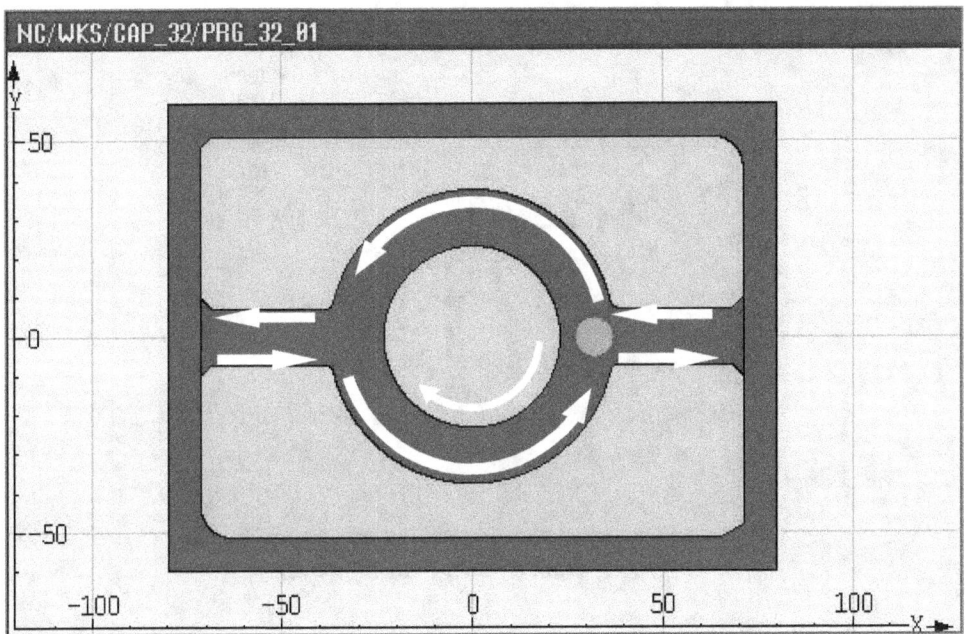

Fig. 150. Finishing of the internal profile

17.10 Program, phase 4: execution of the holes

The holes are programmed by using the drilling cycle and the MCALL function as explained in chapter 23.

```
N510  T="CUTTER 4" D1 M6 ; MILL DIAM. 4
N520  G95 S2300 M3 M8 F0.12
N530  G0 X59.2 Y28.8
N540  G0 Z2

N550  MCALL CYCLE82(10,0,2,-9,,0.6,0,1,12)

N560  G0 X59.2 Y28.8
N570  G0 X48.8 Y28.8
N580  G0 X-48.8 Y39.2
N590  G0 X-48.8 Y28.8
N600  G0 X-48.8 Y-28.8
N610  MCALL

N620  G0 Z500
N630  M30
```

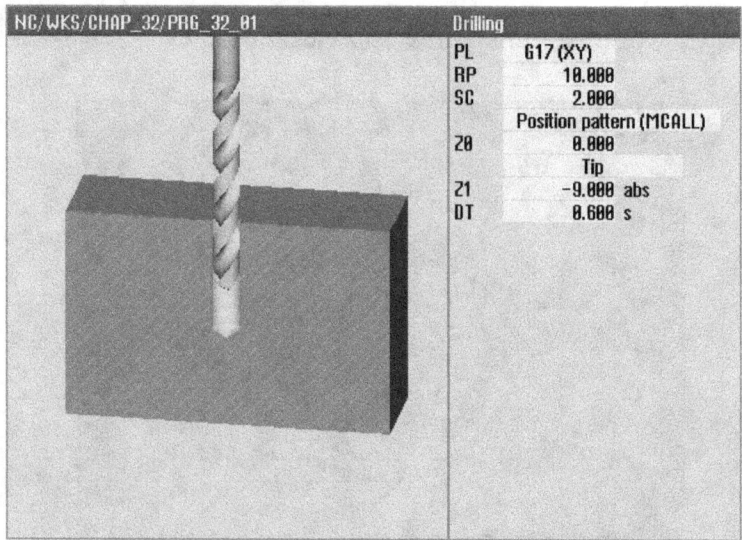

Fig. 151. Data entered in drilling cycle

17.11 Program, phase 5: activation of graphic simulation

Open the program PRG_17_01 in the folder CHAP_32 and activate the graphic simulation. Then execute the program in single block mode and reduce the execution speed by setting the potentiometer in the graphic display to 80%.

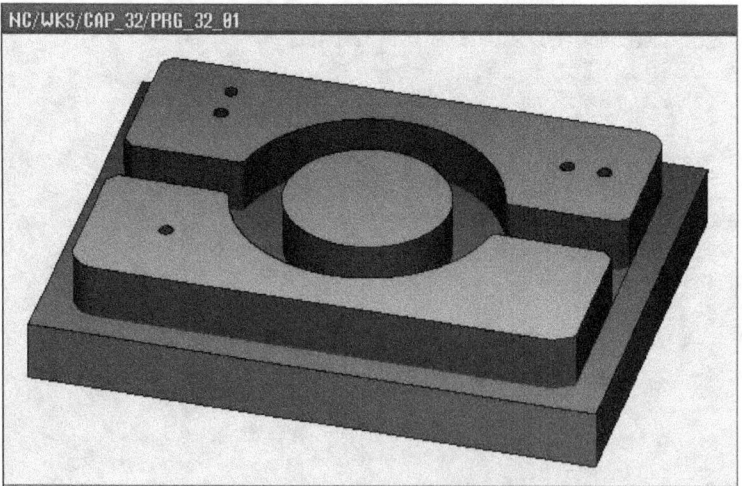

Fig. 152. Graphic 3D image of the finished part

18. Climb Milling and Conventional Milling (2h)
(Theory: 2h)

18.1 Peripheral milling

18.1.1 Introduction
Peripheral milling is performed when the rotating axis of the cutter is parallel to the machining surface, whether it is vertical or horizontal. The feed direction can be either **discordant (up or conventional milling)** or **concordant (down or climb milling)** with respect to the cutting speed vector of the mill, as shown in the following figure. The feed direction is chosen by the operator according to the information shown in the following paragraphs. The cross section of the chip has an increasing profile in peripheral conventional milling and a decreasing profile in peripheral climb milling.

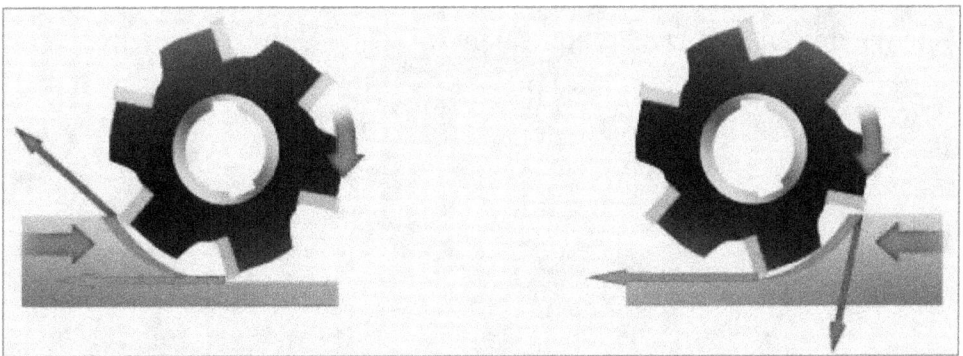

Fig. 153. Discordant cutting direction (left) and concordant cutting direction (right)

18.1.2 Chip section area
The longitudinal section (with respect to the rotating axis) of the chip has the shape of a rectangle both in conventional and in climb milling.

Referring to the following figure, one side of the rectangle is constant and equal to 'b', while the other is variable and equivalent to the thickness 's'. The feed per tooth is indicated with 'a_z'.

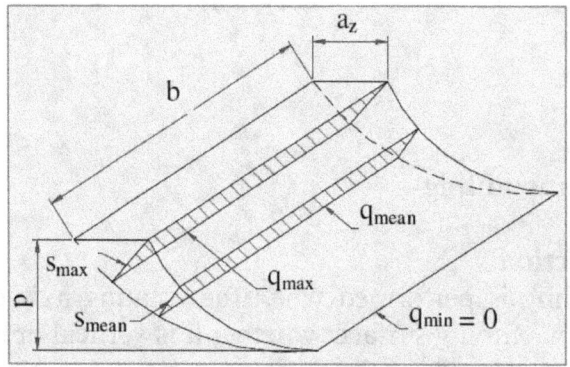

Fig. 154.   Chip section area

The maximum thickness 'S_{max}' is calculated with the formula:

$$S_{MAX} \cong 2 \cdot a_z \cdot \sqrt{\frac{p}{D}} \quad (mm) \quad con \quad \begin{cases} p = & \text{cutting depth} \\ D = & \text{cutter diameter} \end{cases}$$

The maximum area 'q_{max}' is equivalent to:

$$q_{MAX} = b \cdot s_{MAX} \cong b \cdot 2 \cdot a_z \cdot \sqrt{\frac{p}{D}} \quad (mm^2)$$

18.1.3 Conventional milling: cutter and workpiece movement

The feed direction is discordant with respect to the cutting speed vector of the mill.

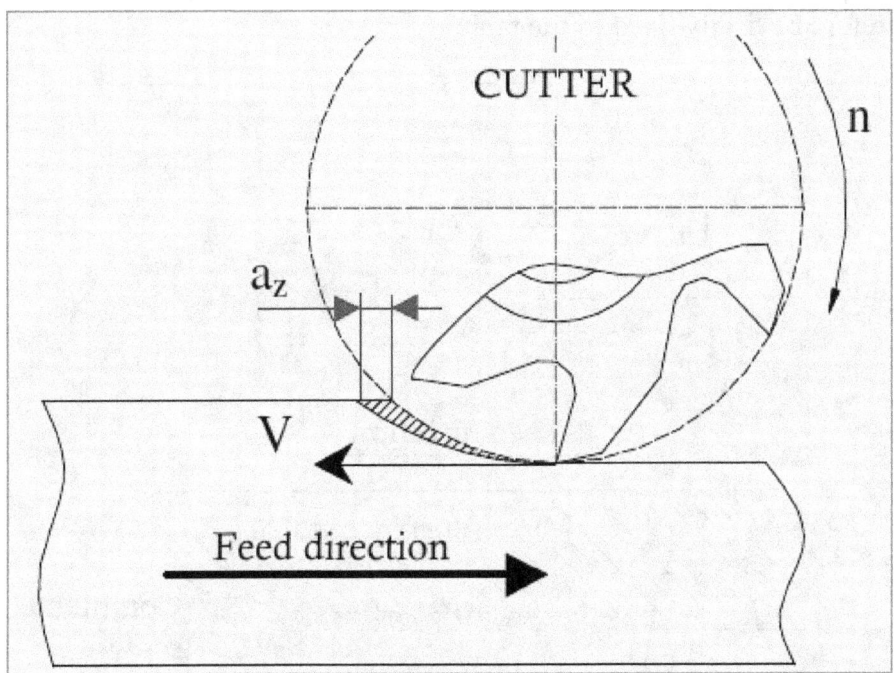

Fig. 155. Relative movement between cutter and workpiece with discordant feed direction

The cross-section (with respect to the rotating axis) of the chip has an increasing profile, varies from zero at the start point of stock removal and ends with a width equal to the feed per tooth 'a_z' at the end point of stock removal. During machining, the cutting edge scrapes against the material; the friction absorbs power and heats up the material causing it to harden. Conventional milling usually results in a poor finish and accelerates the wear of the cutting edges.

18.1.4 Conventional milling: cutting force distribution

The cutting force 'F_t' is tangent to the trajectory performed by the cutting edge of the tooth moving over a cycloid arc. By dividing this force at the point of maximum stress into two vectors, it can be seen that the component of the cutting force parallel to the 'F_o' table goes in the opposite direction with respect to the feed direction; this allows to maintain the contact between the thread flanks in the kinematic mechanism of screw and nut of the axis that makes the table move, neutralizing possible backlashes in the coupling. The component of the cutting force orthogonal to the 'F_v' table tends to lift the workpiece, which must therefore be firmly fixed to the table.

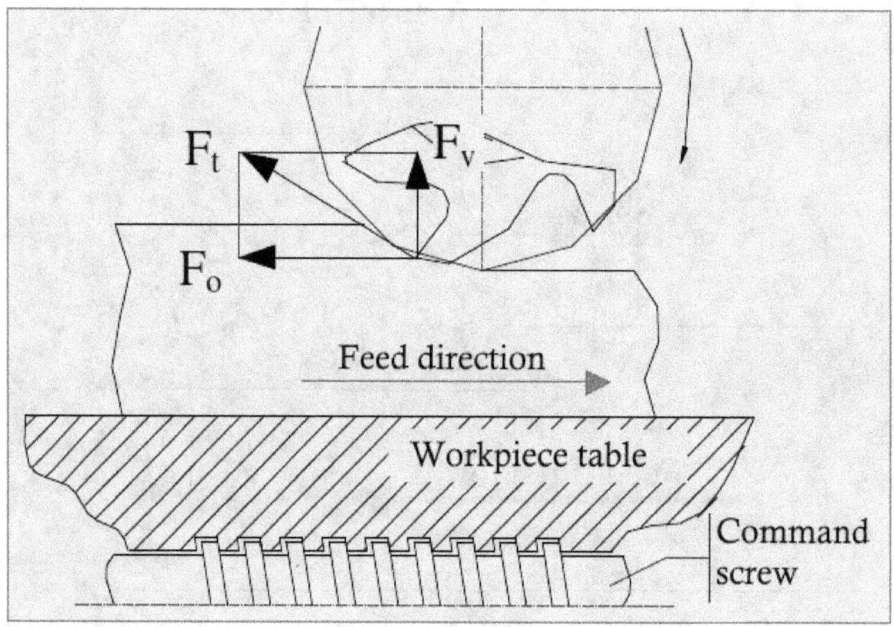

Fig. 156. Cutting forces with discordant feed direction

18.1.5 Climb milling: cutter and workpiece movement
The feed direction is concordant with respect to the cutting speed vector of the mill.

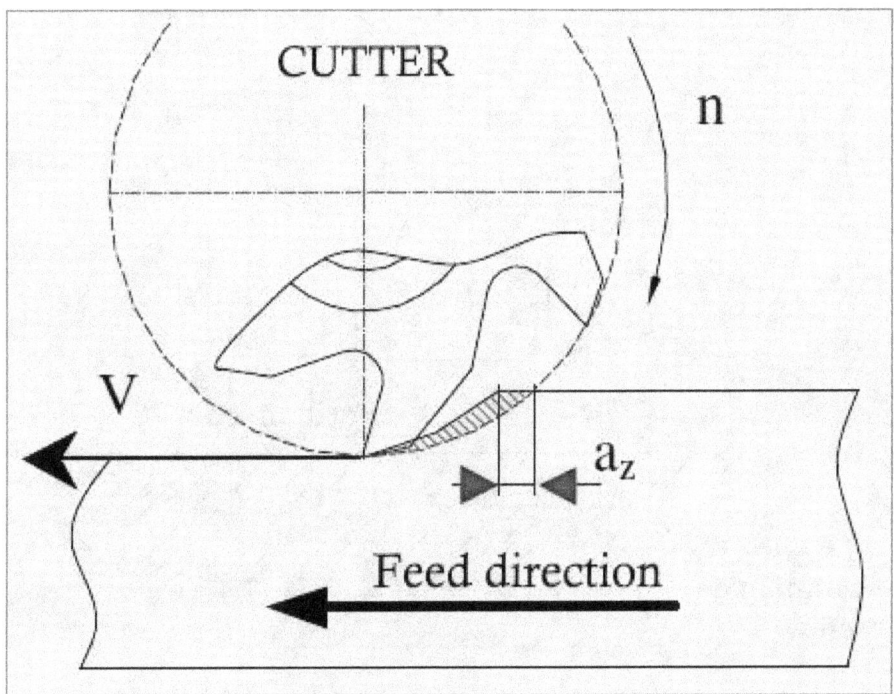

Fig. 157. Relative movement between cutter and workpiece with concordant feed direction

The cross-section (with respect to the rotating axis) of the chip has a decreasing profile, its width at the start point of stock removal is equal to the feed per tooth 'a_z' and it finishes with zero thickness at the end point of stock removal. The produced heat is absorbed mostly by the chip, preventing the material from heating up. Climb milling also facilitates the chip removal.

18.1.6 Climb milling: cutting force distribution
As in the previous case, the cutting force 'Ft' is tangent to the trajectory performed by the cutting edge of the tooth. By dividing this force at the point of maximum stress into two vectors, it can be seen that the component of the cutting force parallel to the table 'F_o' table goes in the same direction with respect to the feed direction; this can cause the screw

thread flanks to detach from the nut if the coupling has play. The component of the cutting force orthogonal to the table 'F_v' table tends to compress the workpiece thereby facilitating the fastening system.

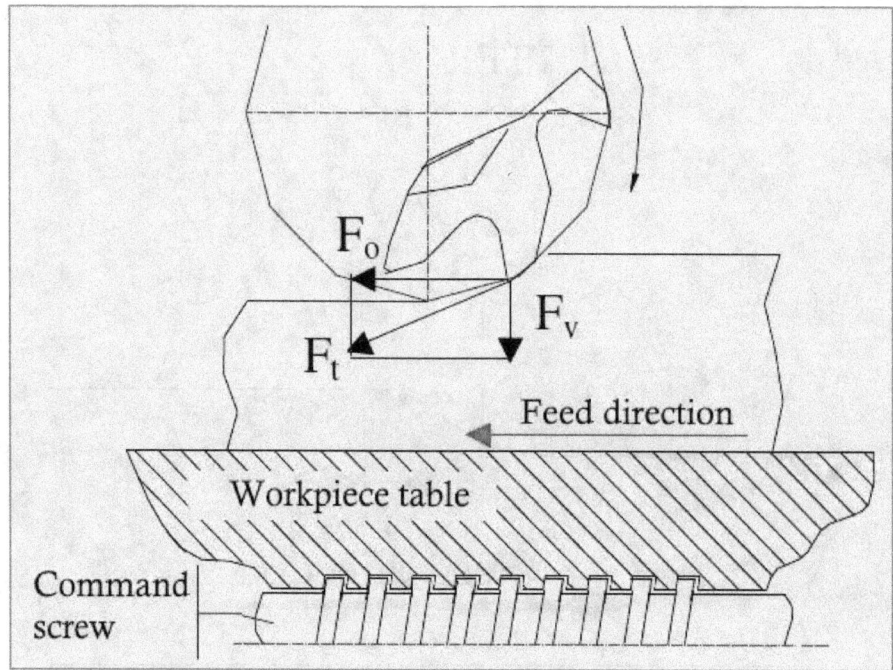

Fig. 158. Cutting forces with concordant feed direction

18.1.7 Conclusions

It can therefore be said that climb milling, as long as it is performed on milling machines with transmission systems with automatic backlash recovery, is preferable to conventional milling because of the lower wear of the cutting edges, the greater stability of the workpiece and the absence of tooth flank sliding on the machined surface. Milling with peripheral cutting, both conventional and climb milling, is characterized by a periodic variation of the chip thickness and consequently of the cutting force; this always causes vibrations that must be taken into account when choosing the cutting parameters.

18.2 Face milling

18.2.1 Introduction
Face milling is performed when the rotating axis of the cutter is perpendicular to the surface to be machined. Several teeth are engaged simultaneously during machining. The chip section between the entry point 'A' and the exit point 'C' has small variations in thickness.

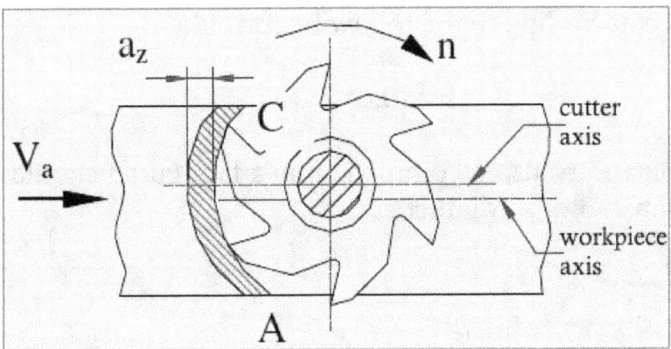

Fig. 159. Cutting forces with concordant feed direction

18.2.2 Chip section area
The chip section is rectangular in shape with sides equal to the pass depth 'p' and the feed per tooth 'a_z'. The section area, which remains almost constant, is calculated using the following formula.

$$q = a_z \times p$$

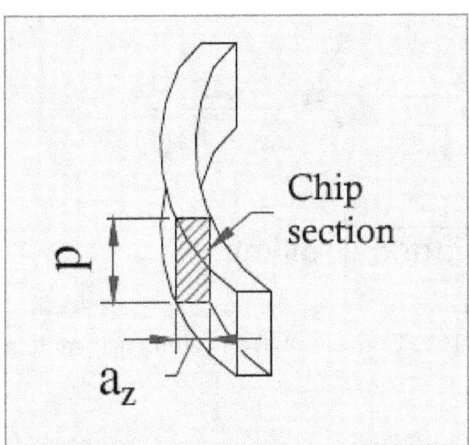

Fig. 160. Cutting forces with concordant feed direction

18.2.3 Climb and conventional face milling

The cutting edge, when working in the arc between points 'A' and 'B', works in concordance; at the same time the cutting edge working in the arc between points 'B' and 'C' works in discordance. When there is play in the coupling between the screw and the nut of the axis, it is possible to avoid the detachment of the thread flanks by decentralizing the position of the cutter by a quantity 's' with respect to the symmetry axis of the pass so that the entry arc 'AB' is greater than the exit arc 'BC'. The value can be calculated according to the following formula.

$$s = (0{,}05 \ / \ 0{,}1) * D$$

Tool manufacturers also recommend that a face cutter should not be used for more than 2/3 of its diameter.

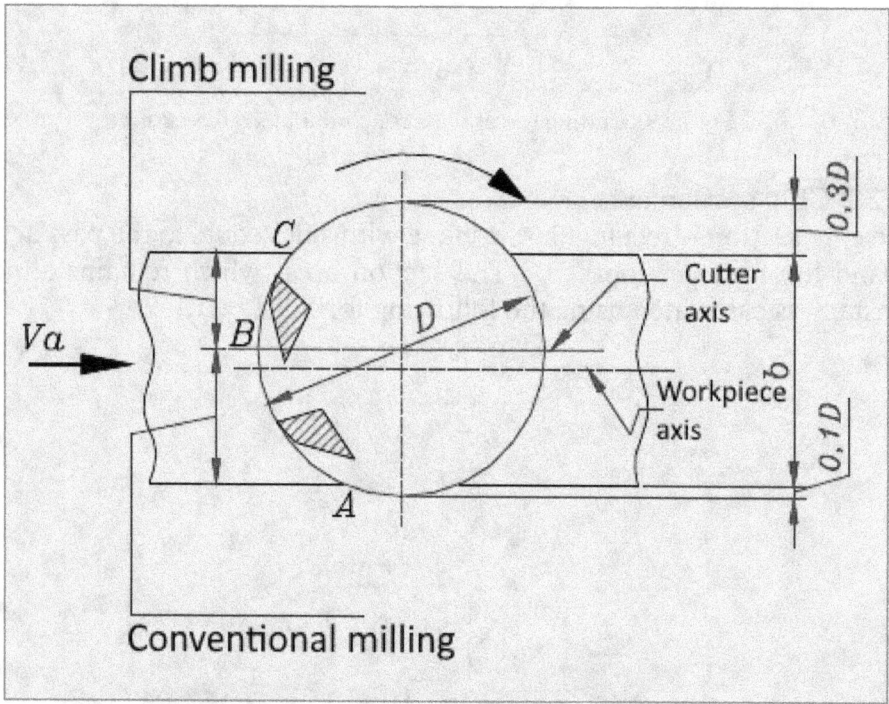

Fig. 161. Conventional and climb face milling

18.3 Comparison of different types of milling

18.3.1 Peripheral conventional milling
In peripheral conventional milling, the cutting edge slides over the material for a short distance before the chip creation begins. The friction generated causes the cutting edge to heat up and its service life is thereby reduced.
The increasing variation in chip thickness ends with an abrupt detachment of the cutting edge from the material causing vibrations that negatively affect the quality of the surface finish.
The direction of the cutting force tends to lift the workpiece from the fastening system.
It should be considered, however, that conventional milling is the only method that can be used in machines with clearance between the screw and nut of the axis on which the machining is performed.

18.3.2 Peripheral climb milling
In peripheral climb milling, the cutting edge begins to cut decisively, immediately encountering a quantity of material to be removed equivalent to the set feed rate.
It generates less vibration, allowing a good surface finish to be achieved.
Most of the heat developed during machining is transmitted to the chip, thereby increasing the service life of the cutting edge.
Under the same cutting conditions, the power required to perform the machining is lower than for conventional milling.
The direction of the cutting force tends to fasten the workpiece to the table.
The only negative aspect of climb milling is that it cannot be used on machines that have clearance in the coupling between the screw and the nut, as this would cause the screw to decouple, increasing the chip thickness by the same amount as the clearance.

18.3.3 Face milling
In this type of milling operation, there are no problems with the selection of the feed direction as long as the cutter works in a way that its axis is properly positioned with respect to the symmetry axis of the pass.
The chip thickness is more constant compared to peripheral milling and the greater number of teeth significantly reduces vibrations, which makes

the machining more uniform and increases the quality of the surface finish.

It is possible to work with higher chip thicknesses because the cutter is usually mounted on a short shaft which reduces tool deflection and allows carbide inserts to be mounted to increase cutting speed.

19. Programming of Four Milled Parts (8h)
(Practice: 8h)

19.1 Programming example with the use of TRANS

The program PRG_19_01 in the folder CHAP_19 creates the part shown in the following figure. It is characterized by the presence of four identical tooling operations executed in four different positions.

The part zero point is located at the center of the rectangle; the program uses the TRANS function to shift it on the X- and Y-axes in the four points from which to start the execution of the profile and the holes.

Fig. 162. Three-dimensional representation of the workpiece

19.1.1 Workpiece program

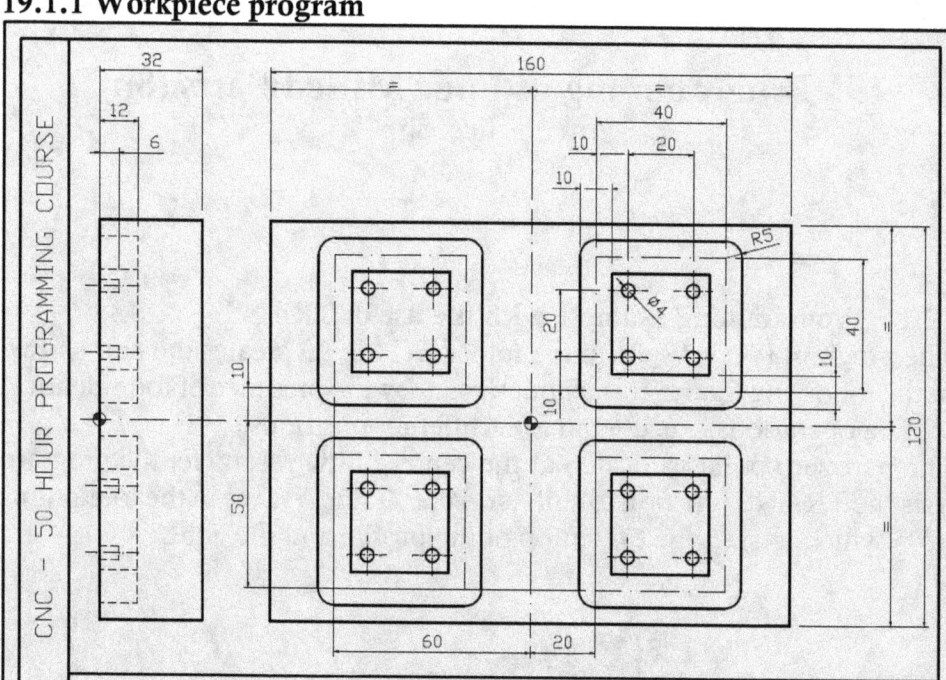

Fig. 163. Drawing of the part to create

```
; blank part: block centered
; W = 120 side length on Y
; L = 160 side length on X
; HA = 0 position of upper face with respect to part zero point
; HI = -32 position of lower face with respect to part zero point

N10 WORKPIECE(,,,"RECTANGLE",64,0,-32,-150,160,120)
N20 G17 G54 G90
N30 G0 Z500

N40 T="CUTTER 10" D1 M6 ; MILL DIAM. 10
N50 G95 S2800 M3 M8
```

N60 TRANS X20 Y10

```
N70 PROFILE1:
N80 G0 Y0 X0
N90 G0 Z2
N100 G1 Z-12 F0.1
N110 G1 X40 F0.18
```

```
N120 G1 Y40
N130 G1 X0
N140 G1 Y0
N150 G1 Z5 F0.8
N160 END1:
```

N170 TRANS X-60 Y10
```
N180 REPEAT PROFILE1 END1
```

N190 TRANS X-60 Y-50
```
N200 REPEAT PROFILE1 END1
```

N210 TRANS X20 Y-50
```
N220 REPEAT PROFILE1 END1

N230 G0 Z500


N240 T="CUTTER 4" D1 M6 ; MILL DIAM. 4
N250 G95 S2300 M3 M8 F0.12
```

N260 TRANS X20 Y10
```
N270 DRILLING1:
N280 G0 X10 Y10
N290 G0 Z2
N300 MCALL CYCLE82(5,0,2,-6,,0.6,0,1,12)
N310 G0 X10 Y10
N320 G0 X30 Y10
N330 G0 X30 Y30
N340 G0 X10 Y30
N350 MCALL
N360 END_DRILLING1:
```

N370 TRANS X-60 Y10
```
N380 REPEAT DRILLING1 END_DRILLING1
```

N390 TRANS X-60 Y-50
```
N400 REPEAT DRILLING1 END_DRILLING1
```

N410 TRANS X20 Y-50
```
N420 REPEAT DRILLING END_DRILLING1

N430 G0 Z500

N440 M30
```

19.2 Example for climb milling

The program PRG_19_02 in the folder CHAP_19 creates the part shown in the following figure. It is characterized by an external and an internal profile.

The workpiece zero point is located at the intersection of the diagonals of the square; the program executes both machining operations with feed in concordance with the rotation direction of the cutter.

You can see that, with right-hand milling cutters, i.e. common standard milling cutters, climb milling is always linked to the tool radius compensation function G41, whether the profile is external or internal.

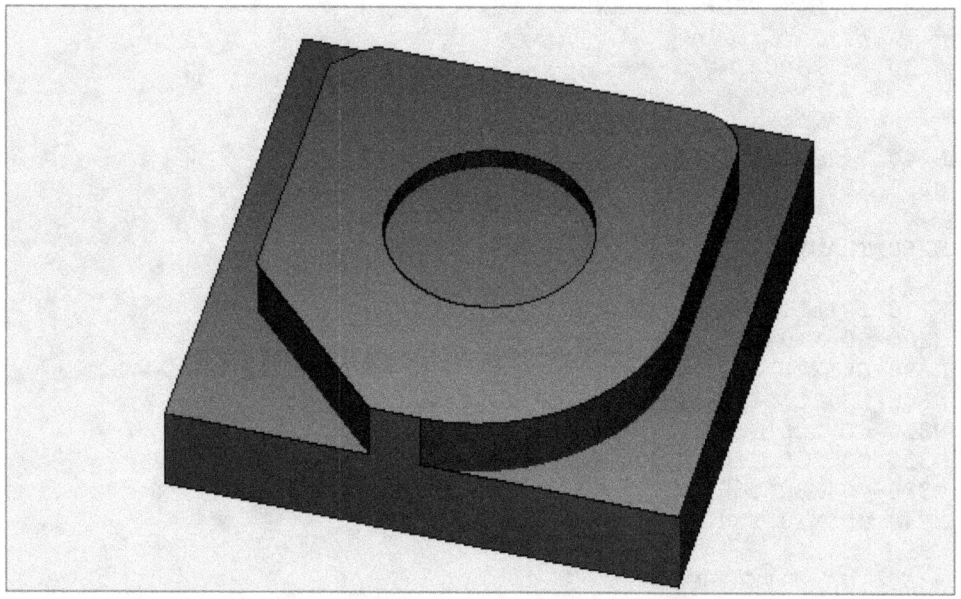

Fig. 164. Three-dimensional representation of the workpiece

19.2.1 Workpiece program

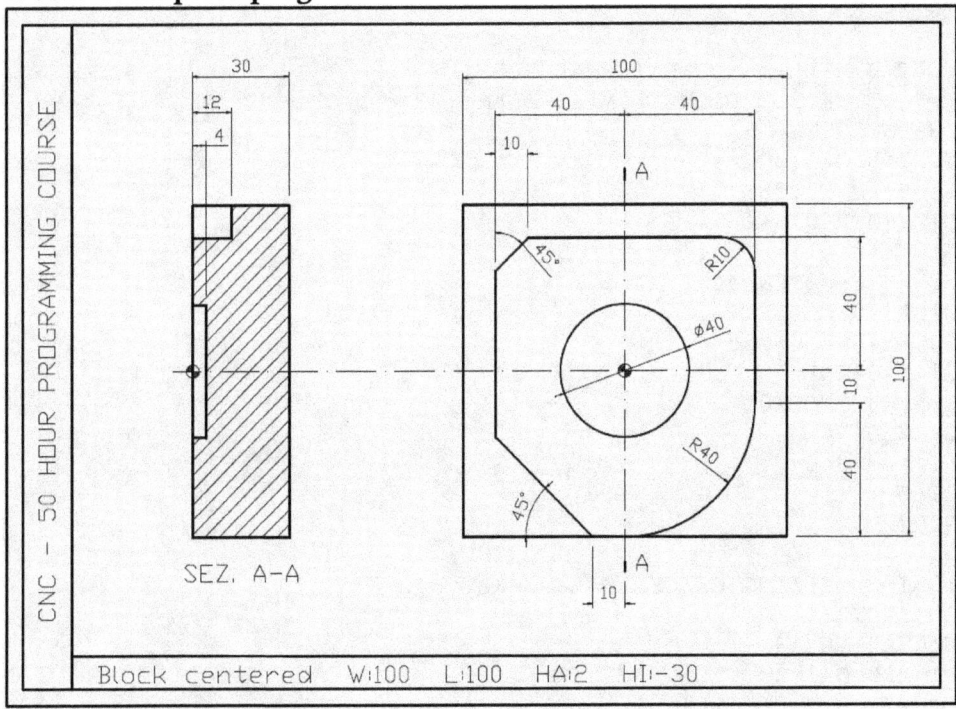

Fig. 165. Drawing of the part to create

```
; blank part: block centered
; W = 100 side length on Y
; L = 100 side length on X
; HA = 0 position of upper face with respect to part zero point
; HI = -32 position of lower face with respect to part zero point

WORKPIECE(,,,"RECTANGLE",64,2,-30,-150,100,100)
G17 G54 G90
G0 Z500
M8

;FLATTENING
T="FACEMILL 63" D1 M6 ; MILL DIAM. 63
G95 S850 M3

G0 Y-30 X85 Z2
G1 Z0 F0.22
G1 X-52
G1 Y30
G1 X52
```

```
G1 Z1
G0 Z500

;CLIMB MILLING OF EXTERNAL PROFILE
T="CUTTER 32" D1 M6 ; MILL DIAM. 32
G95 S850 M3
G0 Y65 X65 Z2
G1 Z-12 F0.18
G1 Y40 X40 G41
G1 Y-10
G2 X0 Y-50 CR=40
G1 X-10
G1 X-40 ANG=135
G1 Y40 CHF=10
G1 X40 RND=10
G1 Y20
G1 X60 G40
G1 Z-10
G0 Z2

;CLIMB MILLING OF CENTRAL POCKET D40
G0 X0 Y0
G1 Z-4 F0.14
G1 X20 G41
G3 X20 Y0 J0 I-20
G1 X0 Y0 G40
G1 Z2

G0 Z500
M30
```

19.3 Programming example using polar coordinates

The program PRG_19_03 in the folder CHAP_19 creates the part shown in the following figure. It is characterized by the presence of an external profile and a series of holes, whose position is defined by a diameter and an angle referred to a center called pole.

The workpiece zero point is located on the lower left edge of the blank part; the program executes the outer profile with a feed discordant to the rotation direction of the cutter and the series of holes using polar coordinate programming.

You can see that, with right-hand milling cutters, i.e. common standard milling cutters, conventional milling is always linked to the tool radius compensation function G42.

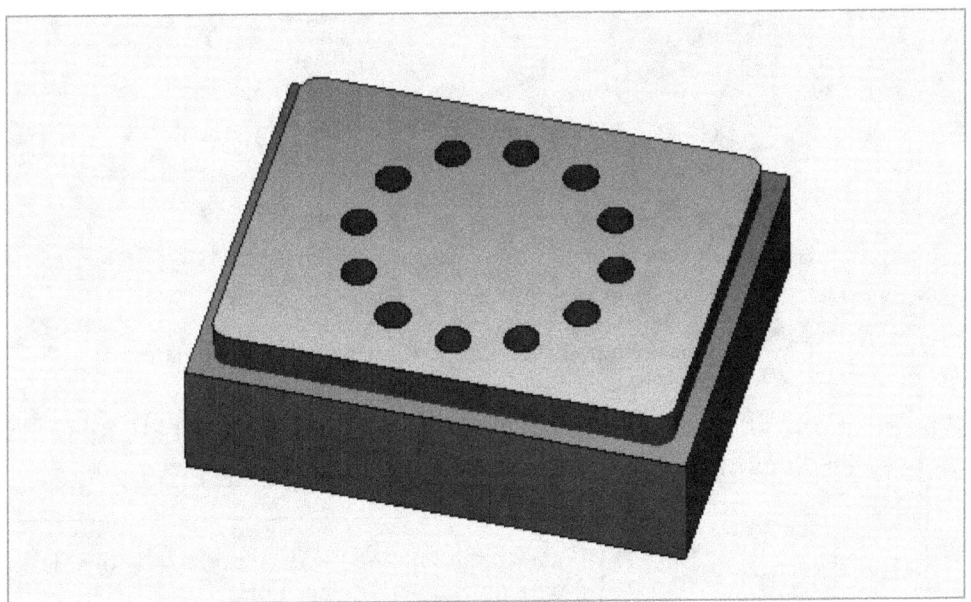

Fig. 166. Three-dimensional representation of the workpiece

19.3.1 Polar coordinates system

The polar coordinate system is a two-dimensional coordinate system in which each point in the plane (in the case of milling, plane G17) is identified by an angle and a distance from a fixed point called pole.

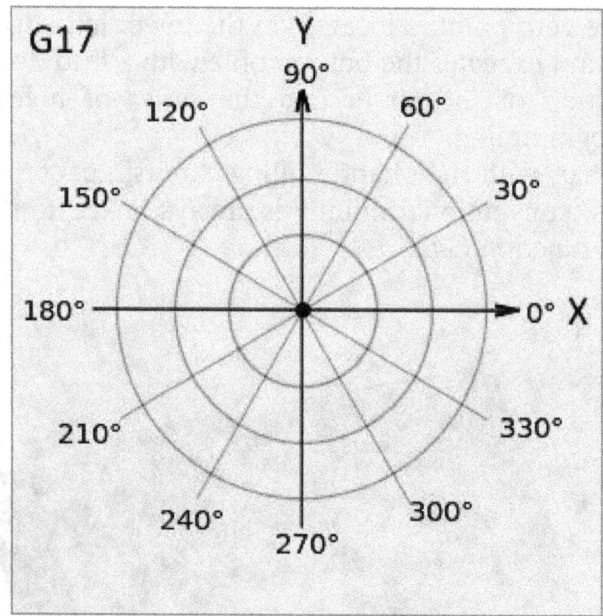

Fig. 167. Definition of a point in plane G17 using polar coordinates

The position of the pole is defined by the functions G110, G111 and G112 as described below and remains valid until the end of the program.

Name	Meaning
G110	Polar programming with respect to the last programmed reference position.
G111	Polar programming with respect to the zero point of the current workpiece coordinate system.
G112	Polar programming with respect to the last valid pole.

If no pole is set, the zero point of the current coordinate system applies.

The position of the point referring to the pole is defined using the functions AP and RP as described below.

Name	Meaning
AP	Polar angle, value range ±0...360°, angle referring to the horizontal axis of the working plane.
RP	Polar radius in mm or inches, always with absolute positive values.

19.3.2 Movement commands with polar coordinates

The point defined by polar coordinates can be reached with a rapid positioning (G0), with a straight working movement (G1) or with a circle arc in clockwise or counterclockwise interpolation (G2 and G3).

```
G0  AP=... RP=...
```

Or

```
G1  AP=... RP=...
```

Or

```
G2  AP=... RP=...
```

Or

```
G3  AP=... RP=...
```

19.3.3 Workpiece program

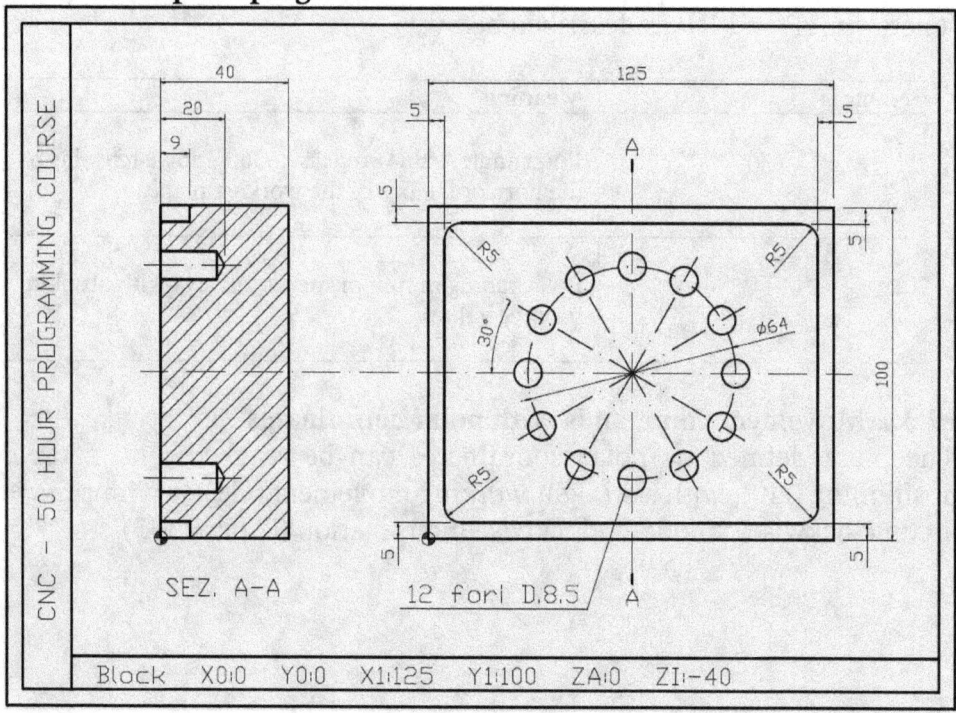

Fig. 168. Drawing of the part to create

```
; blank part: block
; X0 = 0 X coord. of the first corner
; Y0 = 0 Y coord. of the first corner
; X1 = 125 X coord. of the second corner
; Y1 = 100 Y coord. of the second corner
; ZA = 0 position of upper face with respect to part zero point
; ZI = -40 position of lower face with respect to part zero point

WORKPIECE(,,,"BOX",112,0,-40,-150,0,0,125,100)
G17 G54 G90
G0 Z500

;EXECUTION OF EXTERNAL PROFILE BY CONVENTIONAL MILLING

T="CUTTER 16" D1 M6 ; MILL DIAM. 16
G95 S1260 M3 M8
G0 Y-10 X-10 Z2
G1 Z-9 F0.26
G1 Y5 X5 G42
G1 X120 RND=5
```

```
G1 Y95 RND=5
G1 X5 RND=5
G1 Y5 RND=5
G1 X12
G0 Y-10 X-10 G40
G1 Z5 F0.8


;EXECUTION OF THE 12 HOLES USING POLAR COOORDINATES
T="DRILL 8.5" D1 M6 ; DRILL DIAM. 8.5
G95 S1900 M3 M8

G111 X62.5 Y50 ; DEFINITION OF THE POSITION OF THE POLE WITH
RESPECT TO THE WORKPIECE ZERO POINT

G0 RP=32 AP=00 ; HOLE AT 0 DEGREES ON DIAM. 64
G1 Z-20 F0.2
G0 Z2

G0 RP=32 AP=30 ; HOLE AT 30 DEGREES ON DIAM. 64
G1 Z-20 F0.2
G0 Z2

G0 RP=32 AP=60 ; HOLE AT 60 DEGREES ON DIAM. 64
G1 Z-20 F0.2
G0 Z2

G0 RP=32 AP=90 ; HOLE AT 90 DEGREES ON DIAM. 64
G1 Z-20 F0.2
G0 Z2

G0 RP=32 AP=120 ; HOLE AT 120 DEGREES ON DIAM. 64
G1 Z-20 F0.2
G0 Z2

G0 RP=32 AP=150 ; HOLE AT 150 DEGREES ON DIAM. 64
G1 Z-20 F0.2
G0 Z2

G0 RP=32 AP=180 ; HOLE AT 180 DEGREES ON DIAM. 64
G1 Z-20 F0.2
G0 Z2

G0 RP=32 AP=210 ; HOLE AT 210 DEGREES ON DIAM. 64
G1 Z-20 F0.2
G0 Z2

G0 RP=32 AP=240 ; HOLE AT 240 DEGREES ON DIAM. 64
```

```
G1 Z-20 F0.2
G0 Z2

G0 RP=32 AP=270 ; HOLE AT 270 DEGREES ON DIAM. 64
G1 Z-20 F0.2
G0 Z2

G0 RP=32 AP=300 ; HOLE AT 300 DEGREES ON DIAM. 64
G1 Z-20 F0.2
G0 Z2

G0 RP=32 AP=330 ; HOLE AT 330 DEGREES ON DIAM. 64
G1 Z-20 F0.2
G0 Z2

G0 Z500

M30
```

19.4 Programming example with tapped holes

The program PRG_19_04 in the folder CHAP_19 creates the part shown in the following figure. It is characterized by the presence of a series of holes whose position is defined by a diameter and an angle referred to a center called pole.

The workpiece zero point is located at the intersection of the diagonals of the polygon; the program performs the series of holes and the subsequent tapping of each of them using polar coordinate programming and rigid tapping functions.

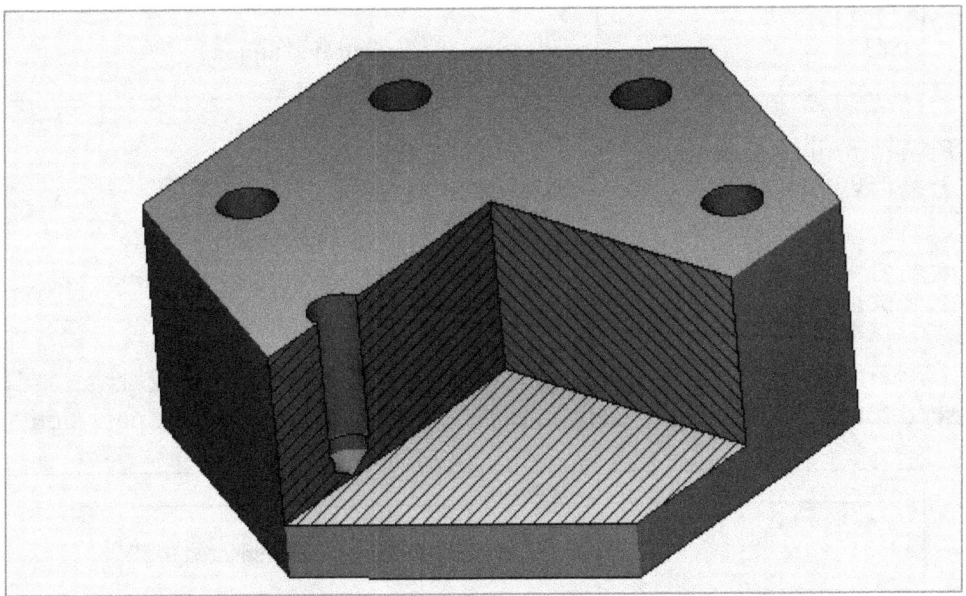

Fig. 169. Three-dimensional representation of the workpiece

19.4.1 Tapping functions

As already seen for the lathe, tapping functions differ for compensated tapping or rigid tapping.

Tapping with a compensated tool is performed when the machine is unable to coordinate the angular position of the tool with its movement along the rotating axis. For this reason, it is necessary to mount the tapping tool on a support that compensates for the axial position error.

The function that activates compensated tapping is G63.

Name	Meaning
G63	Activation of compensated tapping.

Programming example:
```
G95 S600 M3        ; clockwise rotation of the tap
G0 X50 Y50 Z2      ; rapid approach
G63 Z-28 F1 M3     ; tapping
G63 Z2 F1 M4       ; inversion of rotation and return to Z2
G0 Z500            ; disengagement
```

To perform rigid tapping use the functions G331 and G332 as already seen for the lathe. Right-/left-hand threads are defined by the pass sign.

Name	Meaning
G331	Rigid tapping (without compensated tool).
G332	Tool return.

Programming example:
```
SPOS=0                ; angular orientation of spindle to zero
G0 X50 Y50 Z2         ; rapid approach
G331 Z-28 K1 S600     ; tapping
G332 Z2 K1            ; inversion of rotation and return to Z2
G0 Z500               ; disengagement
```

19.4.2 Workpiece program

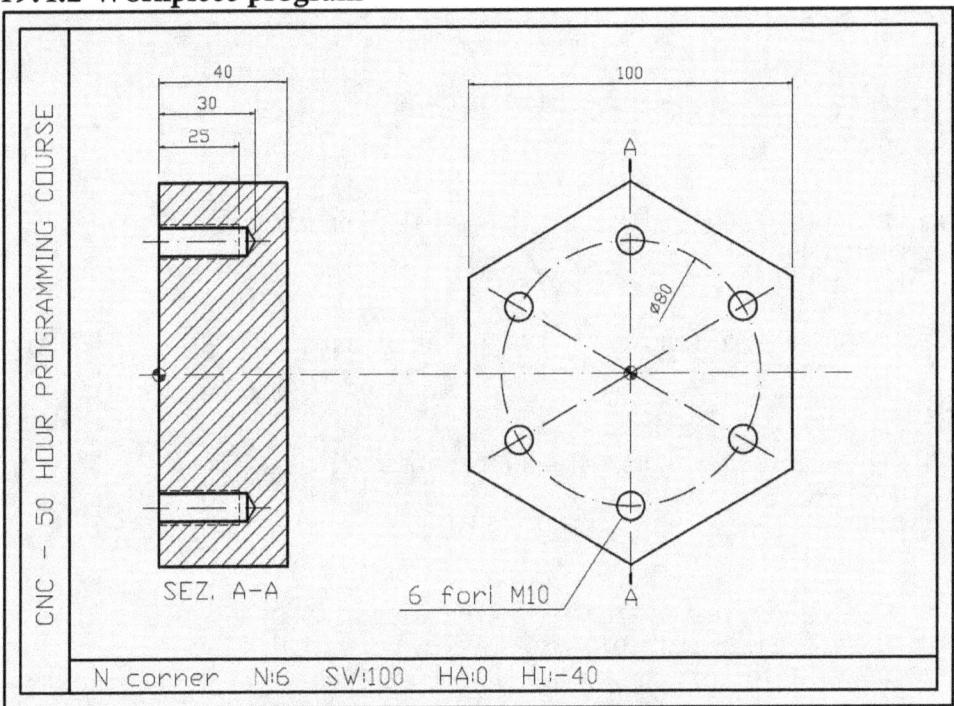

Fig. 170. Drawing of the part to create

```
; blank part: N corner
; N = 6 number of corner
; SW = 100 side to side distance
; HA = 0 position of upper face with respect to part zero point
; HI = -40 position of lower face with respect to part zero point

WORKPIECE(,,,"N_CORNER",64,0,-40,-150,6,100)
G17 G54 G90
G0 Z500

;EXECUTION OF THE 6 HOLES USING POLAR COOORDINATES
T="DRILL 8.5" D1 M6 ; DRILL DIAM. 8.5
G95 S1900 M3 M8

G111 X0 Y0 ; DEFINITION OF THE POSITION OF THE POLE

G0 RP=40 AP=30 ; HOLE AT 30 DEGREES ON DIAM. 80
G1 Z-30 F0.2
G0 Z2
```

```
G0 RP=40 AP=90  ; HOLE AT 90 DEGREES ON DIAM. 80
G1 Z-30 F0.2
G0 Z2

G0 RP=40 AP=150 ; HOLE AT 150 DEGREES ON DIAM. 80
G1 Z-30 F0.2
G0 Z2

G0 RP=40 AP=210 ; HOLE AT 210 DEGREES ON DIAM. 80
G1 Z-20 F0.2
G0 Z2

G0 RP=40 AP=270 ; HOLE AT 270 DEGREES ON DIAM. 80
G1 Z-30 F0.2
G0 Z2

G0 RP=40 AP=330 ; HOLE AT 330 DEGREES ON DIAM. 80
G1 Z-30 F0.2
G0 Z500


;EXECUTION OF THE 6 RIGID TAPPING OPERATIONS
T="THREADCUTTER M10" D1 M6 ; TAP M10
G95 S680 M3 M8

G111 X0 Y0 ; DEFINITION OF THE POSITION OF THE POLE

G0 RP=40 AP=30 ; TAPPING AT 30 DEGREES ON DIAM. 80
SPOS=0
G331 Z-25 K1.25 S680
G332 Z2 K1.25

G0 RP=40 AP=90 ; TAPPING AT 90 DEGREES ON DIAM. 80
SPOS=0
G331 Z-25 K1.25 S680
G332 Z2 K1.25

G0 RP=40 AP=150 ; TAPPING AT 150 DEGREES ON DIAM. 80
SPOS=0
G331 Z-25 K1.25 S680
G332 Z2 K1.25

G0 RP=40 AP=210 ; TAPPING AT 210 DEGREES ON DIAM. 80
SPOS=0
G331 Z-25 K1.25 S680
G332 Z2 K1.25

G0 RP=40 AP=270 ; TAPPING AT 270 DEGREES ON DIAM. 80
```

```
SPOS=0
G331 Z-25 K1.25 S680
G332 Z2 K1.25

G0 RP=40 AP=330 ; TAPPING AT 330 DEGREES ON DIAM. 80
SPOS=0
G331 Z-25 K1.25 S680
G332 Z2 K1.25

G0 Z500

M30
```

20. Second Test (2h)
(Pratice: 2h)

20.1 Introduction to the test
The test consists in the execution of the program creating the part shown in figure 173. Take the following steps:
- Load the tool files which you find in folder 01_EXERCISES named EMPTY_TOOL_LIST. This file deletes all the existing tools by overwriting them only with the end mill defined therein.
- Now create the tools you need to run this program. Below is a list of the necessary tools, their position in the magazine, the offset data on the Z-axis and the data for the definition of their graphic aspect.

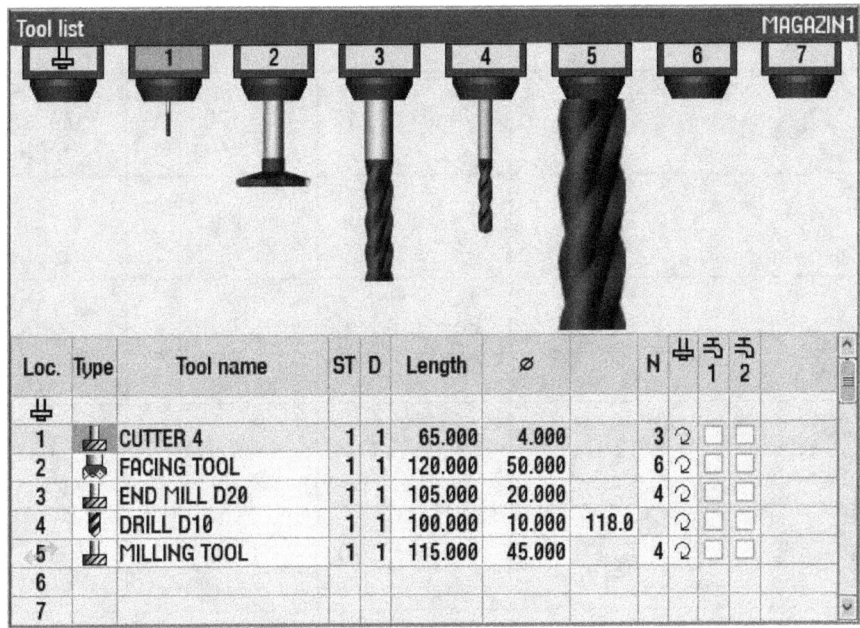

Fig. 171. List of tools to be created and used in the test program

- In the folder 01_EXERCISES, create an empty main program and name it TEST_20_01.
- Structure the program as the ones we've seen so far:
- Insert at the top of the program the comments that show the dimensions of the blank part.
- Define the dimensions of the blank part.
- Enter the blocks that activate the initial settings and the home position:
  G17 G54 G90
  G0 Z500
- Proceed to the programming of the tooling operations following the logical sequence described in paragraph 5.2.

20.2 Tooling operations and cutting parameters

Tooling sequence	Tool name	Operation	Cutting speed (m/min)	Feed rate (mm/rev)
1st	T2 D1	Flattening	100	0.6
2nd	T5 D1	Cylinder D112	120	0.3 climb milling
3rd	T3 D1	Pocket D80	90	0.2 climb milling
4th	T3 D1	Four milling operations	110	0.32
5th	T1 D1	N2 holes D4	80	0.06
6th	T4 D1	N4 holes D10	80	0.12

Fig. 172. Sequence of tooling operations and cutting parameters to use for the test

20.3 Drawing of the part to create

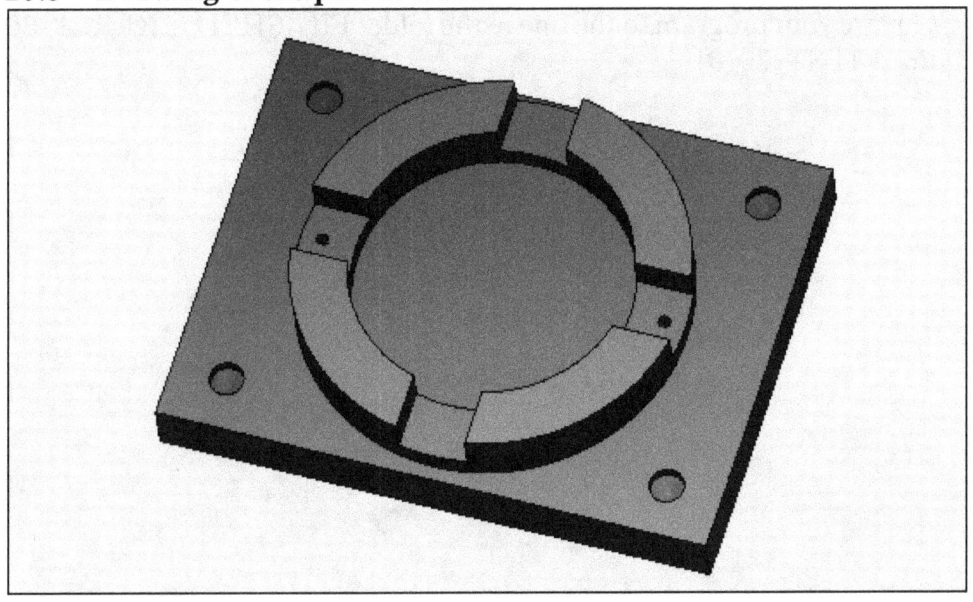

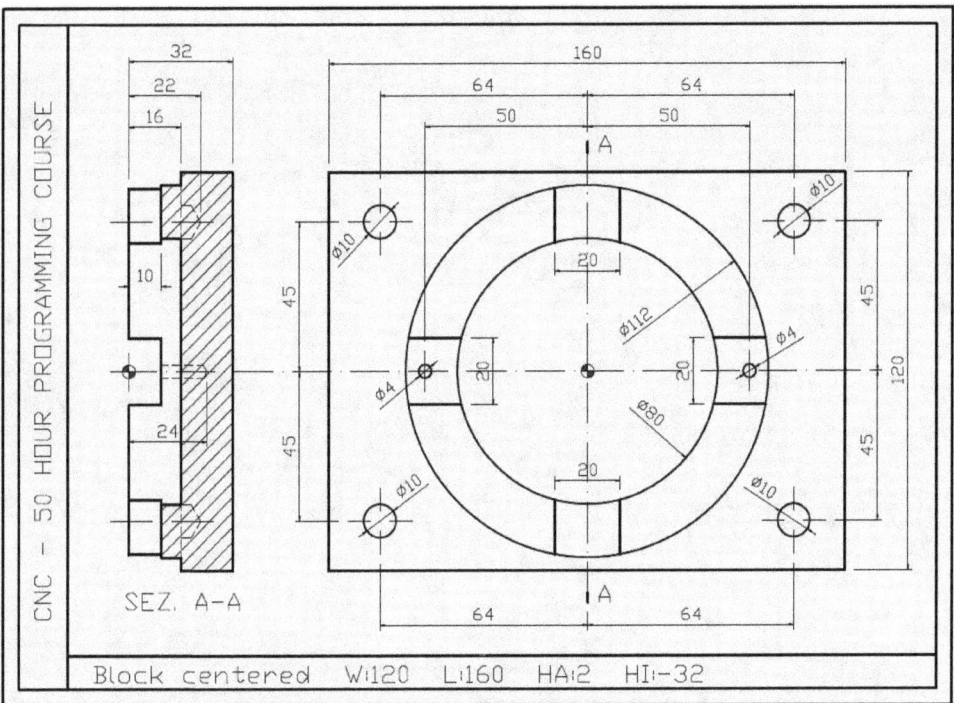

Fig. 173. Drawing of the part to create

20.4 Program correction
Compare your program to the one in the folder FINISHED_EXERCISES named TEST_20_01.